FLORA OF TROPICAL EAST AFRICA

SAPOTACEAE

J. H. Hemsley

Trees or shrubs, rarely climbers, generally with milky juice. Leaves alternate, simple, always entire in Africa. Stipules present, often caducous, or absent. Flowers solitary or clustered in the axils or at nodes below, ⚥ or rarely ♀ by reduction of stamens, regular, generally small. Calyx with 4–8 sepals or shortly united lobes in one or two whorls, rarely spiral. Corolla usually cream or white, campanulate to shortly tubular, with 4–8 lobes in 1–2 series and sometimes divided into 3 segments. Stamens as many as the corolla-lobes and opposite them or more numerous and in 2(–several) whorls; staminodes sometimes present between the corolla-lobes, variously developed; anthers 2–thecous, opening lengthwise. Ovary superior, usually 5–many-locular; style simple; ovules solitary in each locule and ascending from the inner angle. Fruit a berry, with a generally thin outer layer and a juicy or mealy (rarely tough and leathery) pulp in which the seeds are embedded, rarely a capsule. Seeds with a generally hard smooth often shiny testa; attachment area (scar) small or large, sometimes covering more than half the surface area, softer, often rough and duller in colour; endosperm either copious on either side of flat foliaceous cotyledons or scanty to absent, with the cotyledons then usually thick and fleshy.

A medium-sized family of some 300 species widespread in tropical and subtropical regions.

Authorities differ widely in their concept of the genera, evident in the extensive synonymy given in this account. An accepted basis for classification is found in characters of the calyx, whether uniseriate, biseriate or spirally arranged sepals, the corolla-lobes, whether simple or divided, the development of staminodes, the number of stamens and the variation in number, relative size and fusion of these parts. The composition of the berry, the structure of the seeds (particularly the position, size and texture of the attachment area or scar) and the embryo (the form of which is ±correlated with endosperm development) also provide important characters. The relationship between this promising array of characters is nonetheless markedly reticulate, so that the distinction between many genera is often reduced to a single character and, further, particularly in the number of flower-parts and development of staminodes or endosperm, varies in some parts of the family between apparently closely related species and even within a single species. There has been a general trend to subdivide the larger genera adopted by earlier investigators, to express more naturally the pattern of repeated radiation, perhaps from rather few basic types, which has occurred in all the main regions. It does seem possible, in fact, to define rather satisfactory natural groups on a ± continental basis, with a reasonable correlation of vegetative, flower and fruit characters, but all too frequently these regional concepts break down entirely when applied on a world-wide basis. The most promising solution does, however, seem to lie in a synthesis of such regional groups, employing a variety of characters from all parts of the plant, as has been urged by Professor H. J. Lam in Rec. Trav. Bot. Néerl. 36: 509–525, particularly 514–516 (1939).

This approach, at regional level, has been adopted here and differs in rather numerous respects from the notably divergent systems conveniently summarized recently by Professor Aubréville, Sapotacées, in Adansonia, mém. 1 (1965) and by the late Professor Baehni, Mémoire sur les Sapotacées III, Inventaire des genres, in Boissiera 11 (1965), both evolved after many years of detailed research on a world-wide basis. The system proposed by Professor Aubréville continues the distinguished line of research stemming from the highly original and lucid work of Pierre, only part of which was published in Notes Botaniques Sapotacées (1890–91) and elaborated successively by Baillon (1890–2), Dubard (1911–15), Lecomte (1918–30) and also, in the last three decades, by Aubréville and Pellegrin in Africa and by various workers elsewhere. Variation of flower, fruit and seed characters is analysed in detail and generic rank accorded to any significant divergence, the concept of infrageneric subdivisions, it is argued, obscuring the principles of a binomial classification. Species variable as to the presence of a diagnostic character are placed according to the statistical probability of its presence (" un type dominant, un type statistique "). The African genera so defined do seem largely to form natural, even if often very small groups, some of which others have indeed regarded as sections or subgenera, but where flower and seed characters are labile, even apparently very natural groups, with intermediate forms and coherent in vegetative structure, seem excessively subdivided; for example, the three aggregate species attributed by the present author to *Bequaertiodendron* are distributed between five genera and three tribes. The keys provided require not only both flowering and fruiting specimens but in some cases a wide selection of these, in order to establish the " typical " form of variable parts. The system proposed by Baehni attaches primary importance to seed structure, and species with even minor variation become widely separated; much significance is also attached to the presence of staminodes and even their slight and irregular development in some African species of *Chrysophyllum*, for instance, has caused the dissemination of closely related species. In other respects also, the system, at least as it applies to African representatives, seems rather manifestly artificial.

Thus the nomenclature used in Africa, well established in Engler's detailed treatment in Engler, Monographieen Afrikanischer Pflanzen-Familien und -Gattungen [E.M.] 8, Sapotaceae (1904), has subsequently been in considerable flux and is by no means stabilized even now. The vast increase in material available for study has formed the basis for considerable advance, but there are still all too many gaps in our knowledge. Collectors may assist greatly by seeking both flowering and fruiting material of rare species, and making selective gatherings of all species to show the variations in development of leaves and flowers and the differences between individuals and populations.

The flower-structure in some genera is quite complex and difficulty may be experienced at first in identifying the parts emanating from the corolla-tube. In genera 1–8 and 10 of this account, there is a single series of corolla-lobes, each opposite a stamen,* with or without staminodes of various shape developed inside and between the corolla-lobes. In the other genera, the stamens are each opposed by three petaloid structures, the nature and derivation of which have been the subject of much speculation and varied interpretation. In this account the corolla-lobes are regarded as being divided into three segments, two lateral segments and a single median segment inside and strictly opposite the stamen. The lateral segments correspond with the lateral appendages, appendages, dorsal segments, dorsal appendages or exterior staminodes, and median segments with corolla-lobes, petals or corolla-segments of other authors. This concept is used purely as a working terminology and claims no phylogenetic significance. Staminodes, variously developed, may also occur inside and between the groups of corolla-segments.

Species of three exotic genera have been grown in Tanganyika at Amani. *Palaquium gutta* (Hook.f.) Baill. (*Greenway* 2288 !) and *Payena leerii* (Teysm. & Binn.) Kurz (*Soleman* in *Herb. Amani* G6200 ! & *Greenway* 2287 !) are both native of SE. Asia and have a latex, known as gutta-percha, which has a variety of uses as a non-elastic rubber; both are described in T.T.C.L. *Pouteria campechiana* (H.B.K.) Baehni var. *salicifolia* (H.B.K.) Baehni (*Lucuma salicifolia* H.B.K.), native of Central America, has edible fruits and has also been tried (*Soleman* in *Herb. Amani* G6198 ! & *Greenway* 2942 !). *Chrysophyllum cainito* L. (star apple), *Mimusops elengi* L., *Manilkara zapota* (L.) van Royen (chicle gum tree) and *M. bidentata* (A. DC.) A. Chev. (balata rubber tree) are also grown in East Africa and noted under their respective genera.

Key to genera based on vegetative and flower characters**

Calyx normally consisting of 5 sepals, arranged in a single whorl, but sometimes slightly spiral with a transitional series of sepal to petal-like form from outermost to innermost members (to p. 4):

* Genera with stamens more numerous than the lobes or in more than one whorl are not recorded from East Africa.

** Alternative key to genera based on vegetative and fruit characters on p. 4.

Stipules absent:
- Leaves not pellucid-punctate; style simple or capitate, usually little exserted or if markedly so (*Afrosersalisia, Sideroxylon*) then leaf venation finely reticulate between lateral nerves:
  - Leaves with very numerous, closely parallel, scarcely prominent lateral nerves and short dense appressed indumentum beneath; staminodes often present, but very variable . . . . . . 2. **Bequaertiodendron**
  - Leaves with well-spaced prominent or inconspicuous lateral nerves, or if rather numerous and ± parallel then with the indumentum sparse or lacking:
    - Staminodes normally absent and if present irregular in number and occurrence . 1. **Chrysophyllum**
    - Staminodes regularly present in all flowers:
      - Sepals basally united into short calyx-tube; corolla-lobes of mature flowers spreading at right angles to tube or reflexed; staminodes very small . . . . . . 8. **Afrosersalisia**
      - Sepals ± free to base; corolla-lobes ascending:
        - Staminodes petaloid or elongate and ± ligulate; flowers numerous at each node on pedicels (in East Africa) 1–3 times as long as calyx . . . . . . 5. **Sideroxylon**
        - Staminodes small, subulate; flowers 1–3 at each node on pedicels 4–8 times as long as calyx . . **Imperfectly known genus** (p. 73)
- Leaves (especially thinner textured or epicormic leaves) pellucid-punctate when viewed against a strong light, with prominent well-spaced lateral nerves linked by oblique rather prominent and well-spaced tertiary veins; style slightly expanded into a 5-papillate stigma, short- to long-exserted:
  - Primary lateral nerves looped at and actually forming the leaf-margin; flowers sessile; staminodes absent . 3. **Malacantha**
  - Primary lateral nerves looping near and running closely parallel to leaf-margin; flowers pedicellate; staminodes present. 4. **Aningeria**

Stipules present, at least at the base of young leaves:
- Leaves with very numerous, closely parallel, scarcely prominent lateral nerves and short dense appressed indumentum beneath . . . . 2. **Bequaertiodendron**

Leaves with widely spaced lateral nerves, not closely parallel:
Corolla-lobes entire; stipules persistent, ± as long as young petioles; style exserted in older flowers, rather long:
Flowers subsessile or if pedicellate then pedicels short and robust, not exceeding length of calyx; corolla shortly tubular with spreading lobes ± masking ovary . . . . . 6. **Pachystela**
Flowers with long slender pedicels exceeding length of calyx; corolla-tube very short or absent, with lobes reflexed, clearly revealing ovary. . 7. **Vincentella**
Corolla-lobes divided into 3 segments; stipules weakly developed, generally caducous, very much shorter than young petioles; style short, not exserted . . . . 9. **Inhambanella**
Calyx biseriate, with 6–8 sepals arranged in two distinct whorls, the inner series paler in colour and a little smaller in size:
Leaves with very long petiole, usually one-third to half length of lamina; short shoots stout (to 1 cm. thick) with conspicuous cycad-like scarring . . . . 10. **Butyrospermum**
Leaves and shoots not as above:
Calyx of 8 sepals arranged in two dissimilar whorls of four:
Leaves not crowded at the shoot apices, or if rarely so (*Mimusops schliebenii*) then leaves glossy with open reticulate venation above, rather coriaceous; staminodes linear to lanceolate . . . 11. **Mimusops**
Leaves clustered at the shoot apices, mat with very fine reticulate venation above, ± chartaceous; staminodes broadly ovate . . . . 12. **Vitellariopsis**
Calyx of 6 sepals arranged in two dissimilar whorls of three . . . . . 13. **Manilkara**

KEY TO GENERA BASED ON VEGETATIVE AND FRUIT CHARACTERS*

Calyx normally of 5 sepals, arranged in a single whorl, persistent or deciduous at fruiting stage, but sometimes (9, *Inhambanella*) spirally arranged with 1–2 transitional tepals inside outer series (to p. 6):
Stipules absent:
Calyx not accrescent, membranous, with the lobes free or only shortly united:

* Fruits unknown in the imperfectly known genus described on p. 73, and also in *Afrosersalisia kassneri*, *Pachystela subverticillata* and *Vitellariopsis cuneata*, the generic position of which may in any case require reconsideration when fruits become available.

Seed-scar lateral, long and narrow or broad and covering considerable surface-area; fruits usually distinctly fleshy or coriaceous about the seeds; leaves either with lateral nerves prominent or else striate with many fine closely parallel nerves, rarely coriaceous:

Leaves not pellucid-punctate; style short, simple, rarely persisting on fruits; seed-scar practically smooth:

Fruit (1–)several-seeded, with a fleshy or subcoriaceous pericarp; seeds laterally compressed, with a narrow linear-oblong scar (in East Africa); leaves either without the lateral nerves very numerous and closely arranged or else indumentum sparse to absent beneath . . 1. **Chrysophyllum**

Fruit 1–seeded, with a soft juicy pericarp; seeds oblong-ellipsoid, with a ± broad oblong scar; leaves with very many closely parallel inconspicuous lateral nerves and short very dense appressed indumentum beneath . . . . . . 2. **Bequaertiodendron**

Leaves (especially thinner-textured and epicormic leaves) pellucid-punctate when viewed against a strong light; style persistent on young fruits, usually several mm. long, minutely 5-papillate; seed-scar rugose or tuberculate:

Primary lateral nerves looping at and actually forming leaf-margin; fruits sessile; seed-scar narrowly oblong, covering little surface-area . . . . 3. **Malacantha**

Primary lateral nerves looping near and running closely parallel to leaf-margin; fruits pedicellate; seed-scar oblong-elliptic or elliptic and covering up to half surface-area . 4. **Aningeria**

Seed-scar basal, orbicular or elliptic; fruit small, with a thin fleshy mesocarp; leaves with fine reticulate venation and primary lateral nerves neither prominent nor closely arranged, often coriaceous . . . . . . 5. **Sideroxylon**

Calyx (in East Africa) accrescent, forming a rather woody persistent only slightly lobed cupule at fruit-base; seed-scar covering more than half surface-area . 8. **Afrosersalisia**

Stipules present, at least at the base of young leaves:

Leaves with very many, closely parallel, scarcely prominent lateral nerves and short very dense appressed indumentum beneath . . . . **2. Bequaertiodendron**

Leaves with prominent well-spaced lateral nerves and ± reticulate venation between, glabrous, with scattered hairs or very short indumentum beneath:

Fruits subsessile or on short thick pedicels no longer than calyx; sepals often shortly united at base (fruiting sepals of *P. subverticillata* unknown, but probably free) . . . . . . **6. Pachystela**

Fruits with pedicels much longer than calyx; sepals free:

Fruit ellipsoid, with soft edible pulp around seed; calyx-lobes membranous, persistent, but spreading or reflexed and withered . . . **7. Vincentella**

Fruit broadly obovoid to subglobose, firm-walled, but fleshy and very milky; calyx-lobes thickened, clasping fruit-base . . . . **9. Inhambanella**

Calyx biseriate, with 6–8 persistent sepals arranged in two distinct whorls, the inner series paler in colour and a little smaller in size:

Seed-scar large, covering one-third to half surface-area; leaves crowded on swollen scarred branch-tips; fruits rather large, ± 4–6 cm. long:

Fruit with sweet fleshy pulp; testa glossy, chestnut-coloured; leaf-lamina 2–3 times as long as petiole . . . . . **10. Butyrospermum**

Fruit dryish and coriaceous; testa dull brown; leaf-lamina 6–10 times as long as petiole. **12. Vitellariopsis**

Seed-scar small or narrow, basi-lateral or basal; leaves not clustered at shoot apices, or if so (*Manilkara mochisia*, *Mimusops schliebenii*) then fruits less than 3 cm. long:

Calyx of 8 sepals arranged in two whorls of four; seed-scar usually practically basal and suborbicular*. . . . . . **11. Mimusops**

Calyx of 6 sepals in two whorls of three; seed-scar usually basi-lateral, narrow and several times longer than broad* . . **13. Manilkara**

* Seed difference seems rather constant in East African representatives but is known to break down occasionally elsewhere.

1. CHRYSOPHYLLUM

L., Sp. Pl.: 192 (1753) & Gen. Pl., ed. 5: 88 (1754); Engl., E. & P. Pf. IV, 1: 147 (1897) & E.M. 8: 39 (1904), pro parte maxima, excl. sect. *Zeyherella*; Cronquist in Bull. Torrey Bot. Club 73: 286 (1946)

Gambeya Pierre in Not. Bot. Sapot.: 61 (1891); Aubrév. in Not. Syst. 16: 247 (1960)

Donella Baill., Hist. Pl. 11: 294 (1891) [genus queried]; Aubrév. in Not. Syst. 16: 247 (1960)

Austrogambeya Aubrév. & Pellegr. in Adansonia 1: 7 (1961)

[*Planchonella* sensu Baehni in Boissiera 11: 68 (1965), pro parte, *non* Pierre]

Trees or shrubs, sometimes scandent. Stipules absent. Leaves petiolate; lamina coriaceous or chartaceous; primary nerves usually prominent, straight or curved, closely parallel or widely spaced and then usually with longitudinal or transverse vein reticulum; lower surface often with sericeous indumentum. Flowers clustered in current or fallen leaf axils, subsessile or pedicellate, ⚥*, normally 5-merous. Sepals 5, ± free to base. Corolla ± equal in length to sepals or longer; lobes usually 5, ± equal in length to or shorter than the cylindrical or campanulate tube. Stamens same number as corolla-lobes and opposite, inserted at throat. Staminodes normally absent, but may be irregularly present. Ovary subglobose to broadly conical, densely pilose, normally 5-locular; ovules solitary with lateral or basi-lateral attachment; style short and stout, ± equal length of ovary or rarely longer. Fruit a subglobose to ovoid or ellipsoid berry; pericarp fleshy or subcoriaceous. Seeds several or sometimes solitary; testa hard and shiny; scar lateral and narrow or sometimes (and not in Africa) broad and extending over the seed surface; endosperm present; embryo with flattened foliaceous cotyledons; radicle basal.

A fairly large genus in tropical America, about 20 species in tropical Africa and Madagascar and rather more in the rest of the Old World tropics, though the generic limits become particularly critical in those parts.

Two exotic species are recorded from East Africa, both from coastal and near-coastal areas. *C. cainito* L. has been planted at Amani (e.g. *Soleman* in *Herb. Amani* 6197 ! & *Greenway* 1600 ! & 2362 !), Dar es Salaam (e.g. *Omari* 16 !), in Zanzibar (e.g. Sir John Kirk's old garden, *Greenway* 1318 !) and also probably at other places (see T.T.C.L.: 562 (1949) and U.O.P.Z.: 191 (1949)). This tropical American species has an elliptic leaf, usually 10–12 cm. long, the undersurface of which is clothed with an attractive and persistent reddish-brown silky indumentum. The fruit, known as the star apple or caimito, is greenish-yellow or purplish with a delicious soft pulp. It is usually subglobose and up to 12 cm. in diameter.

The other record, based on a single gathering with ripe fruits, *Drummond & Hemsley* 3494 ! is of *C. lanceolatum* (Bl.) DC. var. *stellatocarpon* van Royen (*C. roxburghii* G. Don). The tree was growing in the E. Usambara Mts. of Tanganyika between Pandeni and Longusa, about 1½ km. N. of Pandeni, in a region previously subject to considerable German influence. A widespread species in tropical Asia, the variety is distinguished by its curiously 5-ribbed fruits, star-like in transverse section (see also footnote on page 9). It has not proved possible to trace the origin of the plant in East Africa but it could conceivably be an earlier German introduction through Amani.

NOTE. Considerable difficulty may be experienced in attempting to name leaves of *Chrysophyllum* species derived from sapling or sucker growth. A wide range of leaf-size and colour of indumentum may be observed in some species, e.g. *C. gorungosanum* Engl., where the indumentum may vary from a pale silvery-brown on the under surface of sapling leaves to the deep ferrugineous brown coloration of crown leaves. Whenever possible mature crown leaves should be obtained and the following key is based primarily on characters derived from such material.

* Female flowers have been observed to occur sporadically among several species examined. Anthers are missing or occasionally the stamen is replaced by a reduced staminodal structure.

Leaves with primary lateral nerves (i.e. nerves running to and looping near the leaf-margin) distinct, arcuate-ascending and spaced at distances of not less than 5 mm. apart:

Leaf-base markedly asymmetrical . . . 5. *C. beguei*

Leaf-base not as above, subequal-sided and narrowly to broadly cuneate:

Petioles short, up to 3·5 mm. in length; leaves small, rarely more than 9 cm. long, elliptic to obovate-elliptic, apex rounded or rarely bluntly and shortly acuminate . . 6. *C. bangweolense*

Petioles usually 1–3 cm. long or longer; leaves larger, normally 10 cm. or more in length, ± 3 times as long as broad:

Leaves narrowly elliptic to oblong-elliptic, 1·5–3·5(–5) cm. wide, secondary lateral nerves (i.e. nerves parallel to but smaller than primary nerves and not reaching leaf-margin) present . . . . 7. *C. muerense*

Leaves elliptic, oblong-oblanceolate or obovate-elliptic, 2·5–9(–14) cm. wide, secondary lateral nerves absent:

Under-surface of leaf with dense reddish or purplish-brown velutinous indumentum covering the entire surface even in oldest leaves, hairs of intercostal areas erect and crisped . . 3. *C. perpulchrum*

Under-surface of leaves with fawn, ferruginous or pale brown sericeous indumentum, sometimes shedding in old leaves, hairs of intercostal areas closely appressed:

Mature crown leaves with tawny or pale brown sericeous indumentum, greyish or silvery-grey or shedding and becoming ± glabrous in old leaves; primary lateral nerves (spacing taken over central part of leaf) normally more than 1 cm. and often more than 1·5 cm. apart; fruits ± glabrous or with remnants of pale tawny indumentum when mature . . . . . 1. *C. albidum*

Mature crown leaves with ferruginous or brownish sericeous persistent indumentum, rarely becoming silvery-grey or ± glabrous in old leaves; primary lateral nerves normally less than 1·5 cm. and frequently less than 1 cm. apart; young fruits with ferruginous indumentum persistent or shedding when mature:

Mature crown leaves 7–15 cm. in length, averaging ± 12 cm.; primary lateral nerves 10–17 in number; fruits relatively small, up to 3 cm. in diameter, usually with persistent ferrugineous indumentum . . . . . 2. *C. gorungosanum*

Mature crown leaves 13–25 cm. long, averaging ± 20 cm.; primary lateral nerves 15–30 in number; fruits large, up to 6 cm. in diameter . . . . . 4. *C. delevoyi*

Leaves with primary lateral nerves numerous and closely parallel, ± straight or slightly curved and spaced at distances less than 5 mm. apart, secondary lateral nerves present and together with the primary nerves giving a striate appearance to the under-surface*:

Leaves large, 7·5–15 cm. long, rarely less than 8 cm. long, oblong-elliptic; fruits large, up to 9 cm. in diameter; seeds up to 5·5 cm. long and 3 cm. wide . . . . . . 8. *C. pentagonocarpum*

Leaves smaller, 4–11 cm. long, usually less than 8 cm. long; fruits smaller, up to 4 cm. in diameter; seeds up to 2·5 cm. long and 1·4 cm. wide:

Leaves drying green, with midrib impressed on upper surface, shortly and bluntly acuminate; fruits subglobose, up to 3·5 cm. in diameter . . . . 9. *C. viridifolium*

Leaves not drying green, without the midrib impressed and usually slightly raised, long and narrowly acuminate; fruits ± ovoid or subglobose, up to 5 cm. long and 4 cm. in diameter** . . . . . . 10. *C. pruniforme*

1. **C. albidum** *G. Don*, Gen. Syst. 4: 32 (1837); Baker in F.T.A. 3: 500 (1877); Engl., E.M. 8: 45, t. 15/C (1904); I.T.U., ed. 2: 391 (1952); F.P.S. 2: 374, fig. 139 (1952); K.T.S.: 524 (1961); Heine in F.W.T.A., ed. 2, 2: 27, fig. 205 (1963); J. H. Hemsl. in K.B. 20: 461 (1966). Type: S. Tomé, *G. Don* (BM, holo.!, K, iso.!)

Tall tree, height up to 45 m., with long straight ± fluted bole and buttressed base. Young shoots, buds and petioles with greyish to tawny-brown indumentum of minute closely appressed hairs. Petioles 1·2–3 cm. in length. Leaf-lamina oblong-elliptic to oblanceolate-elliptic, (8–)12–25(–35) cm. long, 3–9(–10·5) cm. wide, acuminate or sometimes obtuse, narrowly to broadly cuneate; upper surface glabrous, lower surface with soft silky tawny or golden-brown indumentum of minute appressed hairs, later becoming greyish or silvery-grey; midrib and lateral nerves prominently raised, primary lateral nerves (7–)9–14(–16) on each side, arcuate, ascending. Flowers fascicled in current leaf axils or on warty projections on older branchlets. Pedicels 1·5–4 mm. long; both pedicels and calyx with brown pubescence. Sepals broadly ovate to suborbicular, up to 3·5 mm. long, 3 mm. wide. Corolla cream; tube up to 3 mm. long; lobes rounded, up to 2 mm. long,

* Also keying out under this group is a possibly introduced tree, found in the E. Usambara Mts. of Tanganyika, determined as *Chrysophyllum lanceolatum* (Bl.) DC. var. *stellatocarpon* van Royen, distinguished by its prominently ridged subglobose fruits, star-like in transverse section. (See also p. 7.)

** See also *C. sp.* 1 (p. 18).

ciliate. Filaments up to 2 mm. long; small subulate staminodes sometimes present. Ovary up to 2 mm. long, pilose; style up to 2 mm. long. Fruit edible, very shortly and stoutly stalked, yellow to orange when mature, depressed globose, up to 5 cm. in diameter, glabrescent. Seeds shiny brown, ± ellipsoid but straight along one margin, laterally compressed, up to 2·5 cm. long and 1·4 cm. wide.

UGANDA. Bunyoro District: Budongo Forest, Busingiro area, May 1933 (fl.), *Eggeling* 1219 in *F.D.* 1326!; Busoga District: Luka, *Eggeling* 724 in *F.D.* 1139!; Mengo District: Mabira Forest, near Najembe, 30 June 1950 (fl.), *Dawkins* 603!
KENYA. N. Kavirondo District: Kakamega Forest, June 1933 (fl.), *Dale* in *F.D.* 3085!
DISTR. **U**1–4; **K**5; widely distributed from West Africa through to the Sudan Republic and Uganda, its easternmost limit appearing to be in the Kakamega Forest, Kenya
HAB. A dominant canopy species of lowland rain-forest and sometimes in riverine forest; sometimes protected and even planted for its fruit; 900–1700 m.

SYN. *Achras sericea* Schumach., Beskr. Guin. Pl.: 179 (1827), *non Chrysophyllum sericeum* A. DC. (1844). Type: Ghana, *Thonning* (C, holo., K, photo.!)
Chrysophyllum millenianum Engl., E.M. 8: 44 (1904). Type: Nigeria, Lagos, *Millen* 47 (B, holo. †, K, iso.!)
C. kayei S. Moore in J.B. 47: 412 (1909). Type: Uganda, Mengo District, Mabira Forest, *E. Brown* 473 (BM, holo.!)
Gambeya albida (G. Don) Aubrév. & Pellegr. in Not. Syst. 16: 247 (1960); Fl. Cameroun 2: 112 (1964)
Planchonella albida (G. Don) Baehni in Boissiera 11: 68 (1965)

NOTE. A common tree in the forests at lower altitudes in Uganda, especially in the Budongo and Mabira Forests. It yields a useful timber and is commercially exploited (see Uganda Forest Dept. Timber Leaflet No. 7). Flowers from Uganda material are a little larger than those from West Africa, with longer corolla-tube in proportion to the lobes. Mature leaves from flowering shoots are correspondingly larger, the shape tending towards oblong-elliptic and with short and bluntly acuminate apices. This is in contrast to leaves of West African material where oblanceolate-elliptic shapes predominate, often with longer acuminate apices. Such variability is considered to fall within the concept of a single widespread species.

2. **C. gorungosanum** *Engl.*, E.M. 8: 44, t. 15/B (1904); Brenan in Mem. N.Y. Bot. Gard. 8: 498 (1954); Meeuse in Bothalia 7: 329 (1960); K.T.S.: 524 (1961); Heine in F.W.T.A., ed. 2, 2: 28 (1963). Type: Mozambique, Manica e Sofala, Mt. Gorongosa [Gorungosa], *Carvalho* (B, holo. †, COI, iso.)

Tall tree, height up to 40 m., with long straight fluted bole. Young branches, buds and petioles with ferrugineous or golden-brown indumentum. Petioles 1–2·5(–4) cm. long. Leaf-lamina narrowly to broadly elliptic, oblanceolate or obovate-elliptic to oblong-elliptic, 7–15(–25) cm. long, 2·3–5·5(–8) cm. wide, obtuse to broadly acute or shortly acuminate, cuneate; lower surface with dense ferrugineous, golden-brown or pale silvery-brown or silvery sericeous indumentum of short closely appressed hairs; midrib and lateral nerves prominently raised, primary lateral nerves 10–17(–24) on each side, arcuate, ascending. Flowers clustered in current leaf axils, shortly pedicellate; pedicels 1–2 mm. long; pedicels and calyx densely ferrugineous pubescent. Sepals broadly ovate, up to 3·5 mm. long, 3 mm. wide. Corolla cream or pale yellowish; tube up to 2 mm. long; lobes rounded, up to 2 mm. long, ciliate. Filaments up to 2 mm. long, flattened; small triangular or ± subulate staminodes occasionally present. Ovary densely pilose; style up to 1·5 mm. long. Fruits very shortly stalked, ± ovoid to subglobose or ellipsoid, up to 4 cm. long, 3 cm. in diameter, with dense reddish-brown pubescence, frequently rubbing away in irregular patches. Seeds dark brown or blackish, obliquely ellipsoid to obovoid, up to 2·8 cm. long and 1·2 cm. wide. Fig. 1.

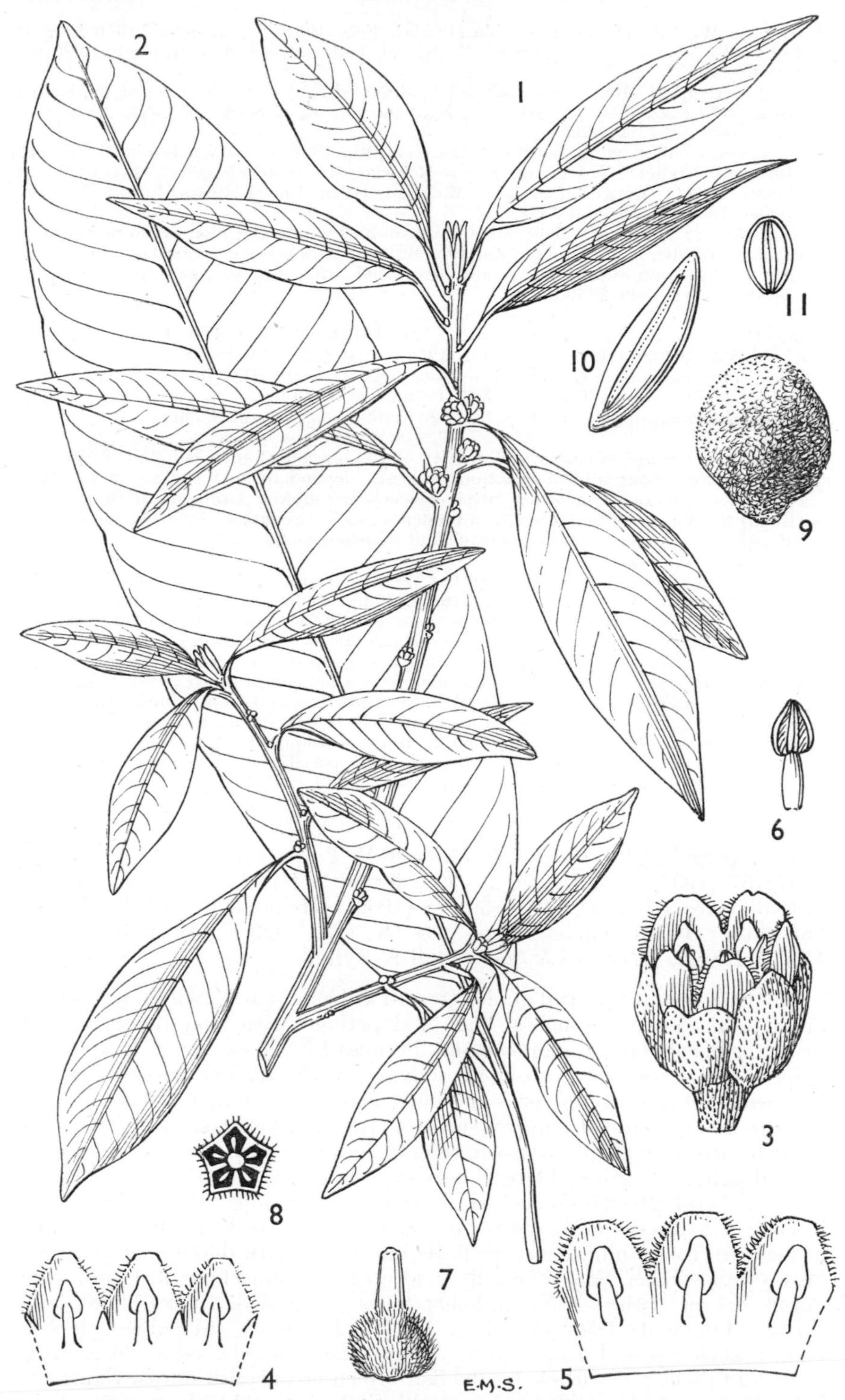

FIG. 1. *CHRYSOPHYLLUM GORUNGOSANUM*—**1,** flowering branch, × $\frac{2}{3}$; **2,** sapling leaf, × $\frac{2}{3}$; **3,** flower, × 6; **4,** section of corolla, with stamens and small staminodes, × 6; **5,** section of corolla lacking staminodes, × 6; **6,** stamen, × 6; **7,** ovary, × 6; **8,** diagrammatic transverse section of ovary; **9,** fruit, × $\frac{2}{3}$; **10,** seed, × 1; **11,** transverse section of seed, × 1. 1, from *Hockliffe* in *F.D.* 1370; 2, from *Drummond & Hemsley* 4364; 3, 5–8, from *Eggeling* 1510 in *F.D.* 1440; 4, from *Battiscombe* 560; 9–11, from *Drummond & Hemsley* 4365.

UGANDA. W. Nile District: Payida [Paida], Feb. 1934 (fl.), *Eggeling* 1510 in *F.D.* 1440!; W. Ankole Forest, *Dawe* 357!; Kigezi, Impenetrable Forest, Sept. 1936 (fl.), *Eggeling* 3302!

KENYA. Aberdare Mts., *Battiscombe* 560!; S. Nyeri/Embu Districts: S. of Mt. Kenya, *Hockliffe* in *F.D.* 1370!; Teita Hills, Ngangao Forest, 15 Sept. 1953 (fr.), *Drummond & Hemsley* 4364! & 4365!

TANGANYIKA. W. Usambara Mts., Magamba, Jan. 1932 (fl.), *Wigg* 106 in *F.H.* 790!; Morogoro District: N. Uluguru Forest Reserve [above Morningside], May 1953 (fr.), *Semsei* 1194!; Songea District: Matengo Hills, Liwiri-Kiteza Forest Reserve, 20 Oct. 1956 (fr.), *Semsei* 2591!

DISTR. **U**1, 2; **K**4, 7; **T**3, 6–8; Mt. Cameroon and mountains along eastern border of Congo Republic, southwards to Zambia, Malawi, Mozambique and Rhodesia

HAB. Upland rain-forest, commonly associated with *Podocarpus* spp. and *Ocotea usambarensis*; 1300–2250 m.

SYN. *C. fulvum* S. Moore in J.L.S. 40: 131 (1911); T.S.K.: 120 (1936); T.T.C.L.: 562 (1949); I.T.U., ed. 2: 392 (1952); F.P.S. 2: 374 (1952). Type: Rhodesia, Chipinga District, Chirinda Forest, *Swynnerton* 19 (BM, holo.!, K, iso.!)
[*C. albidum* sensu T.T.C.L.: 562 (1949), *non* G. Don]
[*C. boivinianum* sensu F.F.N.R.: 320 (1962), *non* (Pierre) Baehni]

VARIATION. A range of colour and density of the lower leaf-surface indumentum may in part be local geographical variation but is also dependent upon leaf-age and position on the tree. Frequently growing with and closely resembling *Aningeria adolfi-friedericii* (Engl.) Robyns & Gilbert, there is a tendency, as in the latter, for evolution of small infraspecific segregates under conditions of montane isolation.

NOTE. The species is closely allied to the West African *C. delevoyi* De Wild. and also the Madagascan *C. boivinianum* (Pierre) Baehni, occupying the median part of the range. Typically a tree of montane forest (cf. the lowland forest habitat of both *C. boivinianum* and *C. delevoyi*), *C. gorungosanum* is distinguished mainly on fruit characters. Fruits of *C. delevoyi* are much larger, up to 7 cm. in diameter, borne on thick woody stalks, while those of *C. boivinianum* are small, up to 2 cm. in diameter, subglobose and densely covered with a purplish-brown felted indumentum. The timber is of good quality but often uneconomical to cut, due to the depth and extent of the fluting. The species is probably more widespread in montane forest than present material would indicate and it may be possible to define segregates more precisely when the range and variation is better understood.

3. **C. perpulchrum** *Hutch. & Dalz.*, F.W.T.A. 2: 10 (1931) & in K.B. 1937: 57 (1937); I.T.U., ed. 2: 395, fig. 80/b (1952); Aubrév., Fl. For. Côte d'Ivoire, ed. 2, 3: 140, t. 30/6–8 (1959); Heine in F.W.T.A., ed. 2, 2: 28 (1963). Types: Ghana, *Vigne* 234 (K, syn.!, OXF, isosyn.!) & *Vigne* 1185 & Nigeria, *Hitchens & Sankey* (all K, syn.!)

Tall tree, height up to 40 m., buttressed at base; bark smooth, pale brown or greyish. Young branches, buds and petioles with very dense reddish-brown or purple-brown indumentum. Petioles 1·5–3·5 cm. long. Leaf-lamina elliptic, elliptic-obovate or oblong-obovate, 8–20(–40) cm. long, 3–6·5(–14) cm. wide, obtuse, acute or sometimes shortly acuminate, base broadly cuneate to ± rounded; upper surface glabrous, lower surface with deep reddish-brown or purplish-brown velvety indumentum of densely arranged crisped hairs; midrib and lateral nerves prominently raised, primary lateral nerves (11–)14–21(–25) on each side, arcuate, ascending. Flowers congested in current leaf axils or at old nodes, subsessile or very shortly pedicellate; pedicels up to 1·5 mm. long; pedicels and calyx with dense deep purplish-brown pubescence. Sepals broadly ovate, up to 4 mm. long, 3–4 mm. wide. Corolla cream; tube ± 2·5 mm. long; lobes ± rounded, up to 1·5 mm. long, ciliate. Filaments 1·5–2 mm. long, flattened. Ovary subglobose, densely pilose; style up to 1·5 mm. long. Fruit almost sessile, subglobose, up to 3·5 cm. in diameter, with dense reddish-brown or purplish-brown velutinous indumentum. Seeds dark brown, ± elliptic but straight along one margin, strongly laterally compressed, 1·5–2 cm. long, 1–1·4 cm. wide.

UGANDA. Bunyoro District: Budongo Forest, July 1936 (fl.), *Sangster* 142! & July 1931 (fl.), *Brasnett* 131! & Bugoma Forest, 25 Oct. 1956 (fr.), *Trask* in *F.D.* 2103!
TANGANYIKA. Lushoto District: Amani–Kwamkoro road, SW. of Amani, 25 July 1953 (fr.), *Drummond & Hemsley* 3451! & Kwamkoro Forest, 2 Mar. 1960, *Bryce* 132! & 133!
DISTR. **U2**, 4 (based on record in I.T.U.); **T3**; also Congo Republic west to Liberia
HAB. Lowland rain-forest; 800–1200 m.

SYN. *C. africanum* A. DC. var. *orientale* Engl. in E.J. 49: 390 (1913). Types: Tanganyika, E. Usambara Mts., Kwamkoro, *Zimmermann* 297* & Bomole, *Braun* 843 & Derema, *Braun* 860 (all EA, isosyn.!) and others
[*C. africanum* sensu T.T.C.L.: 562 (1949), *non* A. DC.]
Gambeya perpulchra (Hutch. & Dalz.) Aubrév. & Pellegr. in Not. Syst. 16: 247 (1960); Fl. Cameroun 2: 116, t. 25/6–8 (1964)

NOTE. A distinctive and attractive large tree with a heavy looking coppery-green crown, common in the Budongo Forest, Uganda, but apparently with a restricted distribution in East Africa. The timber is used in Uganda but is not of outstanding value.

4. **C. delevoyi** *De Wild.*, Pl. Bequaert. 4: 126 (1926); Heine in F.W.T.A., ed. 2, 2: 28 (1963); J. H. Hemsl. in K.B. 20: 461–465 (1966). Type Congo Republic, Katanga, Nzawa, *Delevoy* 415 (BR, holo.!, K, photo.!)

Tall to medium-sized tree, height up to 25 m. or more, with spreading crown; bole fluted and buttressed at base. Young branchlets, buds and petioles with appressed greyish-brown to deep reddish-brown pubescence. Petiole 1·5–3·5 cm. long. Leaf-lamina elliptic to obovate-oblong, 13–25(–30) cm. long, 4·5–8·5(–10) cm. wide, acuminate, cuneate; upper surface glabrous, lower surface with closely appressed pubescence of short straight hairs, colour varying from pale greyish-brown to deep red-brown; midrib and lateral nerves strongly raised, primary nerves (15–)20–25(–30) each side, ascending, looped near margin. Flowers clustered in current leaf axils or on warty projections on older branches; pedicels up to 4·5 mm. long; calyx and pedicels with ferrugineous pubescence. Sepals broadly ovate, up to 4 mm. long and 3 mm. wide. Corolla creamy-white; tube up to 3 mm. long; lobes ± 2 mm. long, apex rounded, densely ciliate. Filaments up to 2·5 mm. long, flattened; small subulate staminodes occasionally present. Ovary subconical, densely pilose; style tapering, up to 3 mm. long. Fruit shortly pedicellate, orange to orange-yellow when mature, ovoid or ± pyriform to subglobose, up to 6 cm. in diameter, sometimes narrowed at each end, subglabrous. Seeds shiny brown, obliquely elliptic, laterally compressed, up to 3·5 cm. long, 2 cm. wide.

UGANDA. Mengo District: near Nansagazi, Nakiza Forest, 24 Jan. 1951 (fl.), *Dawkins* 700!
DISTR. **U4**; widely distributed in West Africa, Cameroun Republic, Gabon, Angola (Cabinda Province) and Congo Republic, becoming less common towards the east and so far recorded with certainty from Uganda only
HAB. An understory and canopy tree of lowland rain-forest, known outside East Africa also in riverine forest; ± 1200–1400 m.

SYN. [*C. africanum* sensu Baker in F.T.A. 3: 500 (1877); Engl., E.M. 8: 43, t. 15/A (1904); F.W.T.A. 2: 10 (1931), ? sensu A. DC. (1844), *nomen dubium*]
[*Gambeya africana* sensu Pierre, Not. Bot. Sapot.: 63 (1891); Aubrév. in Not. Syst. 16: 247 (1960) & Fl. Gabon 1: 133 (1961) & Fl. Cameroun 2: 114, t. 25/1–5 (1964)]
[*Planchonella africana* sensu Baehni in Boissiera 11: 68 (1965)]

* This gathering consists of three sheets, only one of which is genuine *C. perpulchrum*. Two others are mixtures of *C. perpulchrum* and *Aningeria adolfi-friedericii* (Engl.) Robyns & Gilbert subsp. *usambarensis* J. H. Hemsl. It is perhaps significant that the correct sheet alone bears the original *Zimmermann* label.

NOTE. The species may occur in the Kakamega Forest of Kenya (a single leaf having been seen from this locality) and possibly in other lowland forests. Complete gatherings, including mature fruit, are necessary for a more conclusive determination. There is evidence of selection of favoured varieties in West Africa and one may reasonably anticipate the establishment of cultivar strains with a consequent increase in the diversity of fruit characters.

The doubt which surrounds the typification of the familiar name *C. africanum*, and necessitates the adoption of another name, is discussed in K.B. 20: 461–465 (1966).

5. **C. beguei** *Aubrév. & Pellegr.* in Bull. Soc. Bot. Fr. 81: 795 (1935); I.T.U., ed. 2: 392 (1952); Aubrév., Fl. For. Côte d'Ivoire, ed. 2, 3: 142, t. 303/1 (1959); Heine in F.W.T.A., ed. 2, 2: 27 (1963). Type: Ivory Coast, *Aubréville* 1808 (P, holo.)

Tree up to 30 m. high. Young branches, buds and petioles with ferruginеous indumentum of dense, spreading and intertwined hairs, sometimes ± pilose on epicormic shoots. Petiole ± 5–10 mm. long. Leaf-lamina elliptic to oblong, asymmetrical, 9–14(–20) cm. long, 3·5–6·5(–10·5) cm. wide, shortly acuminate, base broadly cuneate to subcordate, very unequal; upper surface practically glabrous except on midrib, lower surface with dense ferruginous pubescence of spreading hairs; midrib and lateral nerves raised, the latter 8–11(–15) on each side, widely spaced, prominently arcuate-ascending and looped near margin. Flowers clustered on warty projections in leaf axils or at old nodes; pedicels up to 5 mm. long. Calyx pubescent; sepals ± free to base, broadly ovate, to 2 mm. long. Corolla-tube ± 1 mm. long; lobes broadly ovate, up to 0·5 mm. long. Filaments short; anthers mucronulate. Ovary hirsute; style glabrous. Fruit coriaceous, subglobose, up to 2·8 cm. long, 3·3 cm. in diameter, glabrous. Seeds shiny brown, oblong and compressed.

UGANDA. Toro District: Semliki Forest, *Dawe* 645!
DISTR. U2; Ivory Coast, Ghana and Congo Republic
HAB. Lowland rain-forest; 750–900 m.

SYN. *Gambeya beguei* (Aubrév. & Pellegr.) Aubrév. & Pellegr. in Not. Syst. 16: 247 (1960)

NOTE. This distinctive species is represented by a single specimen only from East Africa. The leaves, strongly suggestive of a *Pycnanthus* species, are easily distinguished from those of other East African *Chrysophyllum* species by the asymmetric leaf-base coupled with the relatively widely spaced nerves. Details of flower structure have been taken from the original description. Further collecting in the Bwamba Forest area may confirm the presence of a species last collected by Dawe over fifty years ago.

6. **C. bangweolense** *R. E. Fries*, Wiss. Ergebn. Schwed. Rhod.-Kongo-Exped., 1911–1912, 1: 254, fig. 29 (1916); T.T.C.L.: 562 (1949); F.F.N.R.: 320 (1962). Types: N. Zambia, near Lake Bangweulu [Bangweolo], *R. E. Fries* 909 & 909a (UPS, syn.)

Small tree, height up to 7 m., with greyish or brown reticulated bark. Young branches, buds and petioles with ± dense ferruginous indumentum of spreading and intertwined hairs; older branches practically glabrous. Petioles short, 1·5–3·5 mm. long. Leaf-lamina elliptic, broadly elliptic to obovate-elliptic, 3·5–9(–11·2) cm. long, 1·5–4·5(–6) cm. wide, coriaceous, shortly and bluntly acuminate to rounded and emarginate at apex, cuneate; lower surface with rusty-brown pubescence; nerves raised and prominent on upper surface, less so beneath, primary lateral nerves 7–12 on each side, arcuate, ascending and looped near margin, secondary lateral nerves present and well-developed. Flowers clustered mainly in current leaf axils; pedicels 2–3 mm. long, pubescent. Calyx with sparse to medium pubescence; sepals

± free to base, broadly ovate to suborbicular, up to 3 mm. long, ciliate. Corolla-tube ± 1·5 mm. long; lobes up to 2 mm. long, ciliate. Filaments ± 1·5 mm. long, flattened. Ovary subconical, densely pilose, tapering into 1–2 mm. long style. Fruit shortly stalked, pale yellowish-green, depressed subglobose, up to 4·5 cm. in diameter, with thin smooth skin and firm flesh. Seeds shiny brown, ± ellipsoid and laterally compressed, up to 2·1 cm. long, 1·6 cm. wide; scar lateral and extending to base.

TANGANYIKA. Tabora District: about 60 km. NW. of Tabora, Ichemba, 28 June 1949 (fr.), *Wigg* in *F.H.* 2752! & Tabora, Mar. 1946 (fl.), *Hughes* in *F.H.* 1264!; Singida District: Rift Escarpment near Maw Hills, Oct. 1935, *B. D. Burtt* 5263!
DISTR. T1, 4, 5, 7; Zambia, Congo Republic and Angola
HAB. Deciduous woodland; 900–1700 m.

SYN. *C. cacondense* Greves in J.B. 65, Suppl. 2: 72 (1929). Type: Angola, Bié, R. Luassenha, *Gossweiler* 2623 (BM, holo.!)
Austrogambeya bangweolensis (R. E. Fries) Aubrév. & Pellegr. in Adansonia 1: 7 (1960)

7. **C. muerense** *Engl.* in Z.A.E.: 521, t. 71/C–G (1913). Type: E. Congo Republic, Beni, near Muera, *Mildbraed* 2243 (B, holo. †)

Medium to tall tree, height up to 35 m. Young branches, buds and petioles with greyish-brown to tawny-brown indumentum of dense and closely appressed hairs. Petiole 1–2 cm. long. Leaf-lamina narrowly elliptic to oblong-elliptic, 6–16(–20) cm. long, 1·7–3·5(–5) cm. wide, chartaceous to thinly coriaceous, acute to narrowly acuminate, cuneate; upper surface glabrous with nerves scarcely visible, lower surface with silvery-grey or fawn silky indumentum of dense and closely appressed hairs, rubbing away on older leaves; lateral nerves raised, primary lateral nerves 13–20 each side, arcuate, ascending, secondary lateral nerves irregularly present and inconspicuous. Flowers clustered in axils of current leaves; pedicels up to 3 mm. long; both pedicels and calyx with brown appressed pubescence. Sepals ± free to base, broadly ovate, up to 2·5 mm. long and wide. Fruits carried on short thick woody stalks to 6 mm. long, yellow at maturity, subglobose to obovoid, up to 4·5 cm. in diameter. Seed elongate-oblong and slightly flattened, up to 2 cm. long.

UGANDA. Bunyoro District: Budongo Forest, near Busingiro, Sept. 1932 (fr. & fl. buds), *Harris* 138 in *F.D.* 1104!; Toro District: Bwamba, Aug. 1937 (young fr.), *Eggeling* 3385!; Mengo District: Mabira Forest, *Dawe* 1048!
DISTR. U1, 2, 4; and on the eastern border of the Congo Republic
HAB. Lowland rain-forest; 750–1400 m.

SYN. *Chrysophyllum sp. nov.* sensu I.T.U.: 224, fig. 64/c (1940) & ed. 2: 395, fig. 80/c (1952)

NOTE. The name assigned to this species is based on Engler's description of a specimen obtained from the Beni area to the west of the Ruwenzori range. Unfortunately the type is destroyed. The author gives a figure of a leaf with fruit and seeds and from this and the accompanying description there would appear reasonable grounds for regarding it as the same tree occurring in the Uganda forests. The fruits are relatively large with a thin skin and a thick firm flesh and are carried on a curious short swollen woody stalk. The latter, coupled with the long and narrowly elliptic leaves, provides a useful character for identification. Mature flowers have not been seen or described but a dissection of flower-buds from the specimen *Harris* 138 shows a total absence of stamens and suggests a reduced female form of flower. Further collecting especially of flowering and fruiting material is desirable.

8. **C. pentagonocarpum** *Engl. & Krause* in E.J. 49: 387, fig. 2 (1913); Heine in F.W.T.A., ed. 2, 2: 26 (1963). Type: Cameroun Republic, Moloundou [Molundu], *Mildbraed* 4240 (B, holo. †)

Medium to tall tree, height up to 25 m. or more. Young shoots pubescent only in earliest stages; older branches, petioles and leaves practically glabrous. Petiole 5–10 mm. long. Leaf-lamina oblong-elliptic, 7·5–15 cm. long including the acuminate apex, 6 cm. wide, thinly coriaceous, narrowly acuminate, cuneate, slightly asymmetrical, glabrous on both surfaces; lateral nerves very numerous, closely parallel, ± straight and running almost perpendicular to midrib, looped near margin. Flowers small, clustered in current leaf axils; pedicels slender, up to 5 mm. long, glabrous or puberulous. Sepals free to base, suborbicular, up to 2 mm. long, puberulous externally. Corolla tube ± 1 mm. long; lobes ± truncate, up to 1 mm. long, ciliate. Filaments up to 1·5 mm. long, flattened; staminodes absent. Ovary subconical, densely pilose, tapering to a short thick ± 1 mm. long style. Fruit shortly stalked, subglobose, to 9 cm. in diameter. Seeds large, ± ellipsoid, but straight along one margin and laterally compressed, up to 5·5 cm. long, 3 cm. wide and 2 cm. thick; scar lateral on straight margin, narrowly elliptic.

UGANDA. Toro District: Itwara Forest, Dec. 1943 (fr.), *Eggeling* 5488!
DISTR. U2; Ivory Coast, Ghana, Nigeria, Cameroun Republic, Gabon and Congo Republic
HAB. Lowland rain-forest; ± 1500 m.

SYN. [*C. pruniforme* sensu Dale, I.T.U., ed. 2: 395 (1952), pro parte, *non* Engl.]
Donella pentagonocarpa (Engl. & Krause) Aubrév. & Pellegr. in Not. Syst. 16: 248 (1960); Aubrév., Fl. Gabon 1: 141, t. 25/2 (1961)

NOTE. No mature flowers of this species have been seen from East Africa. Leaves are very similar to those of *C. pruniforme* Engl. or *C. viridifolium* Wood & Franks, but the fruits are large and about the size of an orange and contain the largest seeds found in any East African species of *Chrysophyllum*. The tree would appear to be rare and to be confined to a single forest in W. Uganda; material is very scarce and further collecting is much to be desired.

Professor Aubréville, in Adansonia 3: 231 (1962) & Fl. Cameroun 2: 120 (1964), places this species in the synonymy of *Donella ubangiensis* (De Wild.) Aubrév. (*Mimusops ubangiensis* De Wild.). *Mimusops ubangiensis* was described in Miss. Emile Laurent 1: 434, fig. 81 (1907) from seeds alone. Despite the characteristic form of the seeds of *Chrysophyllum pentagonocarpum* among the known species of the genus, the author hesitates to accept the synonymy of the two names on this evidence alone.

9. **C. viridifolium** *Wood & Franks* in Wood, Natal Plants 6, t. 569 (1912); Gerstner in Journ. S. Afr. Bot. 12: 48–49, fig. 3 (1946); Meeuse in Bothalia 7: 328 (1960); K.T.S.: 525 (1961); Meeuse in F.S.A. 26: 35 (1963). Type: South Africa, Natal, near Durban, *Franks* in *Gov. Herb.* 12520 (NH, holo., K, iso.!)

Small to medium tree, height up to 20 m., with fluted trunk and grey bark. Young shoots, buds and young petioles with ferrugineous pubescence; older branches and petioles becoming glabrous. Petioles 4–10 mm. long. Leaf-lamina oblong to oblong-elliptic, 4–9·5 cm. long, 1·5–4·5 cm. wide, thinly coriaceous, apex tapered and acute to acuminate, base broadly cuneate and usually unequal; upper and lower surfaces glabrous, except on midrib beneath; lateral nerves very numerous and closely parallel, ± straight and slightly arcuate, scarcely ascending. Flowers very small and greenish, clustered in axils of current leaves or at old nodes; pedicels slender, 1·5–3·5 mm. long, puberulous. Sepals ± free to base, suborbicular, ± 2 mm. long, externally puberulous. Corolla-tube ± 1·5 mm. long; lobes rounded, ± 1 mm. long, ciliate. Free part of filaments ± 1·5 mm. long, flattened; staminodes absent. Ovary subconical, densely pilose, tapering to a short thick style, ± 1·5 mm. long. Fruit shortly stalked, yellow when ripe, subglobose, up to 3·5 cm. in diameter, glabrous. Seeds shiny brown, ± ellipsoid, but straight along one margin, laterally compressed, up to 1·8 cm. long; curved margin acute and ± keeled; scar lateral on straight margin, narrowly oblanceolate.

KENYA. Fort Hall/Kiambu District: Thika R. [gorge], 3 Mar. 1955 (fr.), *Nicholson* 63!; Kiambu/Nairobi District: Karura Forest, *Gardner* in *Battiscombe* 938!; Nairobi Arboretum, May 1941 (fl.), *Gardner* 1179 in *C.M.* 16904!
DISTR. **K4**; Mozambique, Rhodesia and south to Natal
HAB. Upland dry evergreen forest and riverine forest; 1500–1700 m.

SYN. [*Chrysophyllum pruniforme* sensu T.S.K.: 119 (1936), *non* Engl.]
[*Chrysophyllum welwitschii* sensu Gomes e Sousa, Essen. Flor. Inhambane: 19 (1943), *non* Engl.]
Donella viridifolia (Wood & Franks) Aubrév. & Pellegr. in Not. Syst. 16: 248 (1960)

NOTE. Closely related to *C. pruniforme* Engl., differing only in a few details. The leaves of *C. viridifolium* dry green or greenish-brown, cf. darkening of *C. pruniforme*. The midrib of the leaf upper-surface is sunken and channelled in *C. viridifolium*, slightly raised or flush with the blade in *C. pruniforme*. The leaf-apex is usually long and tapering into a narrow acumen in *C. pruniforme*, while the leaf of *C. viridifolium* has a shorter and blunter acumen. The fruits of *C. viridifolium* are smaller, subglobose, cf. the ovoid to ellipsoid shapes of *C. pruniforme*.

10. **C. pruniforme** *Engl.*, E.M. 8: 42, t. 14/A (1904); Aubrév., Fl. For. Côte d'Ivoire 3: 122, t. 288/2–9 (1936) & ed. 2, 3: 146, t. 303/2–4 (1959); Heine in F.W.T.A., ed. 2, 2: 26 (1963). Type: Gabon, *Klaine* 283 (B, holo. †, K, iso. !)

Small, medium or tall tree, height up to 30 m. Young shoots, buds and younger petioles with ferrugineous pubescence; older branches and petioles glabrescent or puberulous. Petiole 3–7 mm. long. Leaf-lamina elliptic to oblong-elliptic and tapering at each end, 4–11(–13) cm. long including the acuminate apex, 1·8–5(–6·3) cm. wide, subcoriaceous to coriaceous, apex with long narrow acumen, base cuneate and asymmetrical; upper and lower surfaces glabrous except on midrib beneath; lateral nerves very numerous and closely parallel, ± straight or slightly arcuate, ± ascending. Flowers very small and greenish, clustered in current leaf axils or at older nodes, pedicellate; pedicels 2–3·5 mm. long, glabrescent. Sepals ± free to base, suborbicular, up to 2 mm. long, puberulous or glabrescent. Corolla-tube up to 1 mm. long; lobes rounded or ± truncate, up to 1·5 mm. long, ciliate. Free part of filament up to 1·5 mm. long, flattened; staminodes absent. Ovary subconical, pilose, tapering into a short thick style, ± 1·5 mm. long. Fruit shortly stalked, yellowish when ripe, subglobose to ± ovoid, up to 5 cm. long and 4 cm. in diameter, glabrous. Seeds shiny brown, broadly ellipsoid but ± straight along one margin, laterally compressed, up to 2·5 cm. long, 1·4 cm. wide; curved margin narrowly acute; scar lateral, linear and occupying ± complete length of straight margin.

UGANDA. Kigezi District: Ishasha Gorge, 10 Feb. 1945 (fr.), *Greenway & Eggeling* 7103!
DISTR. **U2**;? **T1**; extending from Sierra Leone, Ivory Coast, Dahomey and Nigeria to Cameroun Republic, Gabon and Central African Republic; no specimens have been seen from the Congo Republic where it almost certainly occurs
HAB. Lowland rain-forest; ± 1500 m.

SYN. *Donella pruniformis* (Engl.) Aubrév. & Pellegr. in Not. Syst. 16: 247 (1960), *nom. non rite publ.*; Aubrév., Fl. Gabon 1: 142, t. 25/6–8 (1961) & Fl. Cameroun 2: 121, t. 26/3–5 (1964)

NOTE. Known with certainty only from the extreme SW. corner of Uganda where it is said to be rare in the Ishasha Gorge. *Procter* 692 from **T1**, Bukoba District, Minziro Forest, appears to be this species, but fruits are necessary for confirmation of this additional provincial record. The species is closely related to *C. viridifolium* and differs only in a few details (see note, above). *C. pruniforme* in the sense of I.T.U., ed. 2: 395 (1952) is a combination of the present species and *C. pentagonocarpum*; the description of the fruit appears to refer to the latter although the specimen *Greenway & Eggeling* 7103 is cited.

Imperfectly known species

C. sp. 1

Medium-sized tree, height up to 8 m., with bushy rounded crown and finely reticulate grey bark, sometimes with fluted trunk. Very young shoots and buds with ferrugineous indumentum, otherwise glabrous. Petioles short, up to 3 mm. long, glabrous. Leaf-lamina oblong to ± oblanceolate, tapering at each end, up to 7·5 cm. long and 2·3 cm. wide, coriaceous and glabrous, apex with long narrow acumen, base narrowly cuneate, slightly unequal; lateral nerves numerous and closely parallel, ± straight and ascending, looped near margin, slightly raised on upper surface.

TANGANYIKA. Iringa District: W. Mufindi, Nyumbanitu, 1 Nov. 1947, *Brenan, Greenway & Gilchrist* 8258 !
DISTR. **T**7; known only from one locality
HAB. Upland evergreen forest; ± 2000 m.

NOTE. General aspect and leaf-characters would point to this gathering having affinity with *C. pruniforme* Engl., but differences are to be noted in leaf-shape and venation of the lower surface. Further material, including flowers and fruits, is needed to determine the gathering with any degree of certainty. The collectors give the native name of Mpalama (Kifuagi) and state the tree to be rare in a small evergreen forest patch.

C. zimmermannii *Engl.*, E.M. 8: 44, t. 15/D (1904); T.T.C.L.: 563 (1949). Type: Tanganyika, E. Usambara Mts., near Amani, *Engler* 855 (B, holo. †)

Tree up to 20 m. high. Young branches and petioles ± densely and shortly fuscous pilose. Petioles 1·5 cm. long. Leaf-lamina oblong and equally narrowed at each end to oblanceolate, 20–25(–33) cm. long, 7–9(–10·5) cm. wide, rigidly membranous, apex obtuse or acuminate with 1·5 cm. long tip; lower surface with greyish silky indumentum of minute appressed hairs; primary lateral nerves 20–26 on each side and prominent beneath, spaced 7–15 mm. apart.

NOTE. This little understood species is known only from the original description, the type specimen having been destroyed. A single sheet in the East African Herbarium, *Braun* 289 from Amani, is written up under this name in Zimmermann's own handwriting. Although the author's original description confines itself to vegetative characters only, a fruit and seeds are figured in the accompanying plate. From characters and habitat given, " clearings in evergreen rain-forest ", sapling or sucker leaf-material might be suspected, which is largely supported by Braun's specimen. The fruit depicted could belong to any of several *Chrysophyllum* species and an error has probably been made in associating the fruit with the leaves described.

The status of the species remains obscure but the possibility of its representing a juvenile state of *C. gorungosanum* Engl. or *C. delevoyi* De Wild. cannot be overlooked. Further study of *Chrysophyllum* species of the Amani region is necessary to resolve the problem.

2. BEQUAERTIODENDRON

De Wild. in Rev. Zool. Afr. 7, Suppl. Bot.: 22 (1919) & Pl. Bequaert. 4: 143 (1926); emend. Heine & J. H. Hemsl. in K.B. 14: 306 (1960)

Chrysophyllum L. sect. *Zeyherella* Engl., E.M. 8: 46 (1904)
Pachystela Engl. sect. *Zeyherella* (Engl.) Lecomte in Bull. Mus. Hist. Nat. Paris 25: 193 (1919)
[*Pouteria* sensu Baehni in Candollea 9: 149 (1942), pro parte; Meeuse in Bothalia 7: 332 (1960), pro parte, *non* Aubl.]
Zeyherella (Engl.) Aubrév. & Pellegr. in Bull. Soc. Bot. Fr. 105: 37 (1958)
Boivinella Aubrév. & Pellegr. in Bull. Soc. Bot. Fr. 105: 37 (1958), *non* A. Camus (1925), *nom. illegit.*

Neoboivinella Aubrév. & Pellegr. in Bull. Soc. Bot. Fr. 106: 23 (1959)
Pseudoboivinella Aubrév. & Pellegr. in Not. Syst. 16: 260 (1960)
[*Amorphospermum* sensu Baehni in Boissiera 11: 102 (1965), pro parte, *non* F. v. Muell.]

Shrubs or trees. Stipules present, linear-subulate, or absent. Leaves petiolate; leaf-lamina chartaceous to coriaceous; primary nerves closely spaced, raised but frequently inconspicuous on lower surface, secondary and sometimes tertiary lateral nerves present, closely parallel; veins ± reticulate and inconspicuous. Flowers congested or sometimes solitary in axils of current or fallen leaves, sessile or pedicellate. Sepals 5, ± free to base. Corolla-tube as long as or shorter than the ± ovate erect lobes. Stamens 5; filaments inserted on tube. Staminodes present or absent. Ovary subconical to subglobose, densely pilose, 5-locular; style short and stout, tapering to simple stigma. Fruit a ± ellipsoid berry with persistent style; skin thin; pericarp soft and juicy. Seed solitary (occasionally more than one ovule develops), broadly to narrowly ellipsoid and sometimes slightly flattened; testa thin but horny; scar lateral or ± oblique to basal, or extending over almost whole of surface area; endosperm absent; cotyledons fleshy, plano-convex; radicle basal.

A small African genus of perhaps no more than three or four very variable species, easily distinguished by the highly characteristic leaves.

Professor Aubréville in Adansonia, mém. 1 (1965), has distributed the material between no less than five genera and three different tribes, but the floral and seed characters on which this segregation is made just happen to be extremely labile in this otherwise natural group. The late Professor Baehni, in Boissiera 11 (1965), similarly distributed the species between several but quite different genera.

Bequaertiodendron is reduced to a synonym of *Englerophytum* Krause (1914) by Professor Aubréville in Not. Syst. 16: 252 (1960) and this may well be correct, but as already noted in K.B. 14: 306 (1960) no authentic material of *Englerophytum* exists, the description and illustration show aberrant characters and the genus was described without fruits, so that its status must remain very uncertain.

Stipules absent; flowers solitary or up to three in number at each node . . . . 1. *B. natalense*

Stipules present, linear-subulate, sometimes caducous; flowers fascicled with several to many flowers at each node:

Lower surface of leaf with ± uniform indumentum of golden-brown or ferrugineous hairs, usually more obvious on youngest leaves; lateral nerves present but inconspicuous; flowers distinctly pedicellate . . . . 2. *B.magalismontanum*

Lower surface of leaf with indumentum of soft silvery-grey closely appressed hairs and scattered emergent deep brown T-shaped hairs; lateral nerves raised and easily visible; flowers sessile . . . . 3. *B. oblanceolatum*

1. **B. natalense** (*Sond.*) *Heine & J. H. Hemsl.* in K.B. 14: 308 (1960); K.T.S.: 523 (1961); Meeuse in F.S.A. 26: 39 (1963). Type: South Africa, Natal, near Durban, *Gueinzius* 181 (S, holo., K, iso.!)

Small to medium tree with spreading branches, height to 25 m. Growth increase by repeated subapical branching; shoots slender with crowded subterminal leaves. Young shoots, buds and young leaves with dense deep

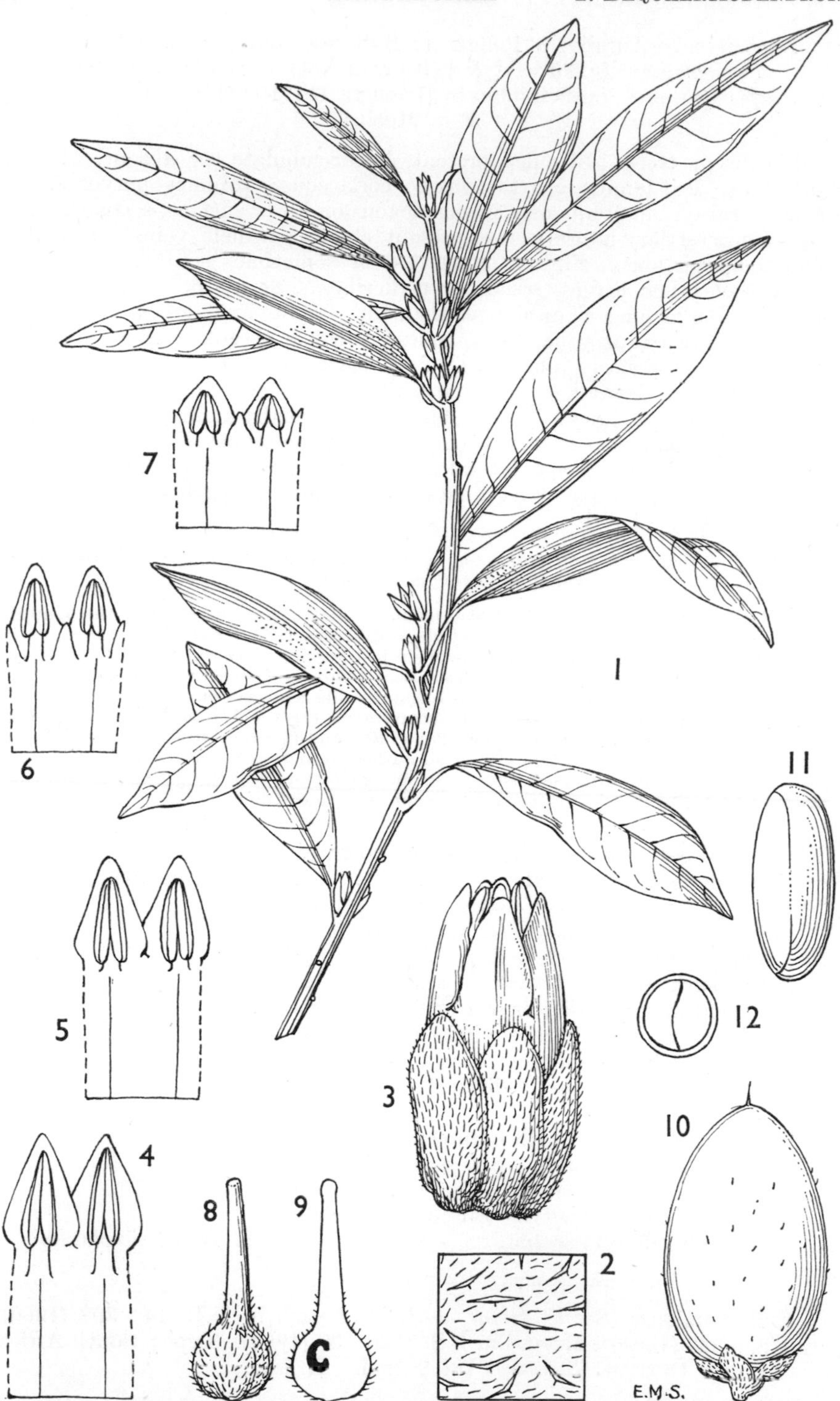

FIG. 2. *BEQUAERTIODENDRON NATALENSE*—**1,** flowering branch, × ⅔; **2,** lower surface of leaf, × 20; **3,** flower, × 6; **4–7,** section of corolla with stamens and staminodes variously developed, × 6; **8,** ovary, × 6; **9,** section of ovary, × 6; **10,** fruit, × 1½; **11,** seed, × 1½; **12,** transverse section of seed, × 1½. 1, 2, from *Eggeling* 6821; 3, 4, 8–12, from *Faulkner* 1080; 5, from *Drummond & Hemsley* 3171; 6, from *Purseglove* 834; 7, from *Eggeling* 3164.

brown indumentum; older branches glabrescent. Stipules absent. Petiole 7–14 mm. long. Leaf-lamina ± oblanceolate to narrowly elliptic, 7–17.5(–20) cm. long, 1·8–5·5(–6·5) cm. wide, chartaceous to thinly coriaceous, obtuse to ± acuminate, narrowly cuneate; upper surface glabrous, lower surface with appressed silvery-grey silky indumentum and with longer scattered brown emergent hairs especially on younger leaves; lateral nerves closely parallel, slightly arcuate, raised but inconspicuous. Flowers solitary or up to three per node, usually in axils of current leaves, subsessile. Calyx with dense dark brown pubescence; sepals ± ovate, up to 5 mm. long, 3·5 mm. wide. Corolla creamy-white; tube up to 4 mm. long; lobes broadly ovate-triangular, up to 2 mm. long, 2·5 mm. wide, ± auricled at base. Free part of filament up to 1·5 mm. long, flattened; anther narrowly obcordate, 1–2 mm. long; staminodes sometimes present, subulate, up to 1 mm. long. Ovary subglobose, ± 1·5 mm. in diameter; style up to 3 mm. long. Mature fruit red, up to 2·5 cm. long, 1·5 cm. in diameter, sometimes with small beak at apex, smooth-skinned. Seed narrowly ellipsoid and blunt at ends, up to 19 mm. long and 8 mm. wide, with large lateral scar occupying most of surface area. Fig. 2.

UGANDA. Ankole District: Lutoto [crater lake], July 1939 (fl.), *Purseglove* 834! & 4 Feb. 1945 (fl.), *Greenway & Eggeling* 7076!
KENYA. Embu District: Kiangombe Hill, 6 Jan. 1955 (fl. & fr.), *Dyson* 416!; Machakos/Masai District: Emali Hill, 15 Mar. 1940, *V. G. van Someren* 109!
TANGANYIKA. Moshi, June 1954 (fl.), *Eggeling* 6821!; W. Usambara Mts., Kwamshemshi–Sakare road, 4 July 1953 (fl.), *Drummond & Hemsley* 3171!; Morogoro District: Nguru Mts., Mtibwa Forest Reserve, Sept. 1952 (fl.), *Semsei* 926!
ZANZIBAR. Zanzibar I., near Haitajwa Hill, 4 Dec. 1930 (fl.), *Greenway* 2651! & Ziwa Chokwe, 5 Aug. 1935, *Vaughan* 2264!
DISTR. **U**2; **K**4, 6; **T**1–3, 5–7; **Z**; and southwards through Malawi, Mozambique and Rhodesia to the Cape Province of South Africa
HAB. Lowland and upland rain-forest, riverine forest and groundwater forest; 0–1800 m.

SYN. *Chrysophyllum natalense* Sond. in Linnaea 23: 72 (1850); Engl., E.M. 8: 43, t. 34/C (1904); Sim, For. Fl. Cape of Good Hope: 252, t. 94 (1909); T.T.C.L.: 563 (1949)
C. kilimandscharicum G. M. Schulze in N.B.G.B. 12: 196 (1934); T.T.C.L.: 563 (1949). Type: Tanganyika, SE. Kilimanjaro, *Schlieben* 4528 (B, holo. †, BM, iso.!)
Boivinella natalensis (Sond.) Aubrév. & Pellegr. in Bull. Soc. Bot. Fr. 105: 37 (1958)
B. kilimandscharica (G. M. Schulze) Aubrév. & Pellegr. in Bull. Soc. Bot. Fr. 105: 37 (1958)
Neoboivinella natalensis (Sond.) Aubrév. & Pellegr. in Bull. Soc. Bot. Fr. 106: 23 (1959)
N. kilimandscharica (G. M. Schulze) Aubrév. & Pellegr. in Bull. Soc. Bot. Fr. 106: 23 (1959)
Pouteria natalensis (Sond.) Meeuse in Bothalia 7: 339 (1960)
Amorphospermum natalensis (Sond.) Baehni in Boissiera 11: 103 (1965)

NOTE. This species appears to be a widely distributed tree in Tanganyika, but becomes much less common in Uganda and Kenya. It is shade tolerant and frequently to be found on banks of streams in shady forest ravines. Dimensions of flower parts increase from north to south and the largest flowers are found in specimens from the type area.

2. **B. magalismontanum** (*Sond.*) *Heine & J. H. Hemsl.* in K.B. 14: 307 (1960); Meeuse in F.S.A. 26: 37, fig. 6/1 (1963); Heine in F.W.T.A., ed. 2, 2: 25 (1963); J. H. Hemsl. in K.B. 20: 469 (1966). Type: South Africa, Transvaal, Magaliesberg, *Zeyher* 1849 (S, holo., K, iso.!)

Shrub or small to medium-sized tree, height up to 33 m., with dense leafy crown and thin fluted bole. Young shoots, buds and young petioles with dense ferrugineous indumentum. Stipules 3–8 mm. long. Petiole 0·4–2·5(–3) cm. long. Leaf-lamina variable in shape, elliptic, elliptic-oblong, elliptic-obovate, oblong-lanceolate or narrowly oblanceolate, 4–19(–24) cm. long, 1·7–5·5(–6·5) cm. wide, coriaceous, obtuse and emarginate or rarely very shortly acuminate, base broadly cuneate to rounded; upper surface of mature leaves glabrous, lower surface of young leaves with soft silky indumentum of dense golden or reddish-brown appressed hairs, sometimes wearing away or changing colour on old leaves, the lower surface then becoming silvery-grey (see note opposite); lateral nerves closely parallel and very numerous, slightly arcuate, barely raised and very inconspicuous. Flowers densely fascicled at nodes on older twigs or on old branches; pedicels 1·5–12 mm. long, densely pubescent with golden-brown or ferrugineous hairs. Sepals broadly ovate to suborbicular, 2–3 mm. long, 2–2·5 mm. wide, densely pubescent externally. Corolla whitish or deep red-brown; tube 1–2·5 mm. long; lobes ± ovate, 2–3·5 mm. long, 2–2·5 mm. wide, auricled at base. Free part of filament 2–4 mm. long, ± terete; anthers 1–3 mm. long; staminodes rarely present and then subulate or broadened and ± ovate, up to 2 mm. long. Ovary subconical; style 1·5–3 mm. long. Fruit red, sometimes slightly oblique or obovoid, up to 2·5 cm. long, 1·8 cm. in diameter. Seed ± ellipsoid and sometimes slightly flattened, up to 1·7 cm. long, 1·4 cm. in diameter; scar lateral to oblique, elliptic to ± narrowly triangular.

TANGANYIKA. Tabora District: near Igalula, 28 Oct. 1934 (fl.), *C. H. N. Jackson* 11!; Mpanda District: 24 km. E. of Uruwira, Busega, July 1951, *Eggeling* 6166!; Lindi District: Rondo, Mchinjiri, Dec. 1951, *Eggeling* 6047!

ZANZIBAR. Pemba I., " Sengenya dume",* 9 Oct. 1951 (fl.), *R. O. Williams* 104!

DISTR. **T**4, 6–8; **P**; widespread from Guinée Republic to Congo Republic and south through Angola, Zambia, Malawi and Mozambique to South Africa (Natal)

HAB. Coastal and deciduous woodlands, riverine vegetation and in ant-hill thickets; 0–1500 m.

SYN. *Chrysophyllum magalismontanum* Sond. in Linnaea 23: 72 (1850); Bak. in F.T.A. 3: 498 (1877); Engl., E.M. 8: 47, t. 16/C (1904); Fl. Pl. S. Afr. 3, t. 98 (1923); Brenan in Mem. N.Y. Bot. Gard. 8: 498 (1954); F.F.N.R.: 321 (1962)

C. argyrophyllum Hiern, Cat. Afr. Pl. Welw. 3: 641 (1898); Engl., E.M. 8: 46, t. 16/B (1904); T.T.C.L.: 562 (1949). Types: Angola, *Welwitsch* 4827 & 4828 (both BM, syn.!, K, isosyn.!) & *Welwitsch* 4829 (BM, syn.!)

C. antunesii Engl. in E.J. 32: 137 (1902). Type: Angola, Huila, *Antunes* 98 (B, holo. †)

Sideroxylon randii S. Moore in J.B. 41: 402 (1903). Type: South Africa, Transvaal, Johannesburg, *Rand* 1017 (BM, holo.!)

Chrysophyllum wilmsii Engl., E.M. 8: 46, t. 16/B (1904). Type: South Africa, Transvaal, Lydenburg, *Wilms* 1812 (B, holo. †, BM, K, iso.!)

C. carvalhoi Engl., E.M. 8: 47 (1904). Type: Mozambique, Manica e Sofala, Mt. Gorongosa, *Carvalho* (B, holo. †, COI, iso.!)

C. mohorense Engl. in E.J. 38: 100 (1905); T.T.C.L.: 563 (1949). Type: Tanganyika, Rufiji District, Mohoro on R. Rufiji, *Gross* in *Herb. Gouvern.* 1030 (B, holo. †)

Pachystela argyrophylla (Hiern) Lecomte in Bull. Mus. Hist. Nat. Paris 25: 192 (1919)

P. magalismontana (Sond.) Lecomte in Bull. Mus. Hist. Nat. Paris 25: 192 (1919)

Chrysophyllum gossweileri De Wild., Pl. Bequaert. 4: 130 (1926). Type: Angola, Benguella, *Gossweiler* 2808 (BR, holo., BM, K, iso.!)

C. lujai De Wild., Pl. Bequaert. 4: 133 (1926). Type: Mozambique, Zambezia, Morrumbala, *Luja* 330 (BR, holo.)

Zeyherella magalismontana (Sond.) Aubrév. & Pellegr. in Bull. Soc. Bot. Fr. 105: 37 (1958) & in Not. Syst. 16: 256 (1960)

Boivinella wilmsii (Engl.) Aubrév. & Pellegr. in Bull. Soc. Bot. Fr. 105: 37 (1958)

B. argyrophylla (Hiern) Aubrév. & Pellegr. in Bull. Soc. Bot. Fr. 105: 37 (1958)

* Not traced; this might be a vernacular name.

Neoboivinella wilmsii (Engl.) Aubrév. & Pellegr. in Bull. Soc. Bot. Fr. 106: 23 (1959)
N. argyrophylla (Hiern.) Aubrév. & Pellegr. in Bull. Soc. Bot. Fr. 106: 23 (1959)
Pouteria magalismontana (Sond.) Meeuse in Bothalia 7: 335 (1960)
Zeyherella gossweileri (De Wild.) Aubrév. & Pellegr. in Not. Syst. 16: 257 (1960)

VARIATION. The inability to distinguish boundaries in a variable range of leaf-size, shape and indumentum and in flower and fruit characters has left no alternative but to consider this taxon as an aggregate species.

Three main tendencies in leaf-character variability are: (*a*) a small-leaved plant from the Transvaal with elliptic to elliptic-oblong or elliptic-obovate shapes and an orange-brown or ferrugineous indumentum on lower surface; (*b*) a larger leaf with narrowly oblanceolate shape ranging from Rhodesia and Mozambique to S. and SE. Tanganyika, the shape sometimes broadening to oblanceolate under favourable conditions, the indumentum being pale in colour and less dense; and (*c*) a larger and elliptic-oblong to elliptic-obovate or oblanceolate leaved group in Zambia, Angola and SE. Congo with the indumentum dense and pale brown or soon becoming silvery grey. There are, however, too many intermediates and exceptions for it to be possible to construct a satisfactory key. Fruit- and seed-size and pedicel-length are very variable and there is a curious occurrence of red to deep purplish-brown flowers in parts of Malawi, Zambia and Angola. The significance of this is not understood. Life-form seems to range from a small gnarled shrub to a medium-sized tree with a long straight trunk or rarely the plant is said to have liane-like branches from the base and to be subscandent.

3. **B. oblanceolatum** (*S. Moore*) *Heine & J. H. Hemsl.* in K.B. 14: 309 (1960); K.T.S.: 523 (1961); Heine in F.W.T.A., ed. 2, 2: 25 (1963). Type: Uganda, Toro District, Durro Forest, *Bagshawe* 1087 (BM, holo.!)

Shrub or much-branched small tree, height up to 10 m.; growth increase by repeated subapical branching; shoots slender with crowded subterminal leaves. Young shoots, buds and young leaves with greyish or brown indumentum; older branches glabrescent. Stipules present, 3–5 mm. long. Petioles 0·7–2 cm. long. Leaf-lamina oblong-lanceolate to oblanceolate, 7–21 cm. long, 2–6·5 cm. wide, chartaceous, obtuse or very shortly acuminate, narrowly cuneate; upper surface glabrous, lower surface with silky greyish indumentum of dense appressed hairs and sometimes with larger brown emergent hairs; lateral nerves closely parallel, slightly arcuate, raised and easily visible. Flowers 5–15, congested in axils of current leaves or at old nodes, subsessile, sweetly scented. Calyx with greyish-brown sericeous indumentum externally; sepals suborbicular, 2–3 mm. long, ± 2·5 mm. wide. Corolla greenish-yellow or pale yellow; tube very short, up to 1 mm. long; lobes broadly ovate, 2–3·3 mm. long, 1·5–2 mm. wide. Filaments free almost to base, up to 4 mm. long, ± terete; anthers up to 1·5 mm. long; staminodes often present, up to 2 mm. long, well-developed subulate structures with incipient lateral processes, or minute and simple. Ovary subconical, up to 2 mm. in diameter; style up to 2 mm. long. Mature fruit red, ellipsoid, up to 2 cm. long, 1·5 cm. in diameter. Seed ± ellipsoid, with widely extending lateral scar.

UGANDA. Toro District: Fort Portal, Kibale Forest, Aug. 1936 (fl.), *Eggeling* 3144!; Bunyoro District, Budongo Forest, Busingiro area, May 1933 (fl.), *Eggeling* 1216 in *F.D.* 1324! & Feb. 1935 (fl.), *Eggeling* 1624 in *F.D.* 1568!

KENYA. N. Kavirondo District: Kakamega Forest, May 1933 (yng. fr.), *Dale* in *F.D.* 3124! & 11 Dec. 1956 (fl.), *Verdcourt* 1680!

DISTR. **U**2; **K**5; Sierra Leone, Ivory Coast, Ghana, Dahomey, Cameroun Republic (no specimen seen) and Central African Republic

HAB. Shrub and lowermost tree layers in lowland rain-forest; 900–1700 m.

SYN. *Sideroxylon oblanceolatum* S. Moore in J.B. 45: 47 (1907)
Chrysophyllum glomeruliferum Hutch. & Dalz., F.W.T.A. 2: 9 (1931) & in K.B. 1937: 56 (1937); I.T.U.: 223, fig. 64/a (1940). Types: Sierra Leone, Mt. Gonkwi, *Scott Elliot* 4867 (K, syn.!) & Dahomey, Savalou, *Chevalier* 23789 (K, syn.!, P, isosyn.!)
[*C. natalense* sensu I.T.U., ed. 2: 392 (1952), *non* Sond.]

Boivinella glomerulifera (Hutch. & Dalz.) Aubrév. & Pellegr. in Bull. Soc. Bot. Fr. 105: 37 (1958)
Neoboivinella glomerulifera (Hutch. & Dalz.) Aubrév. & Pellegr. in Bull. Soc. Bot. Fr. 106: 23 (1959)
Pseudoboivinella oblanceolata (S. Moore) Aubrév. & Pellegr. in Not. Syst. 16: 260 (1960); Fl. Cameroun 2: 101, t. 22 (1964)

NOTE. Said to be a very common undergrowth species in some of the W. Uganda lowland forests, but material, particularly fruiting stages, is still in need of collection. Very similar to *B. natalense* in general appearance, but the easily observable stipules and dense clusters of flowers are good spot characters for differentiating the present species.

3. MALACANTHA

Pierre, Not. Bot. Sapot.: 60 (1891); Baill., Hist. Pl. 11: 295 (1891); Engl., E.M. 8: 47 (1904)

Deciduous tree; young shoots and petioles densely pubescent, older branchlets subglabrous; lenticels pale and raised. Stipules absent. Leaf-lamina large, chartaceous, punctate with small pellucid dots; lateral nerves prominently raised on lower surface, arcuate, looped distally and forming the thickened leaf-margin. Flowers sessile, densely clustered in axils of current or fallen leaves. Buds enclosed within series of hairy bracts. Sepals 5, free to base. Corolla-tube long; lobes 5, rounded, less than half length of tube. Stamens 5; filaments inserted on tube, shorter than corolla; anthers ovate-elliptic. Staminodes absent. Ovary subglobose, densely pilose, 5-locular; style robust, thickened towards apex; stigma 5-papillate. Fruit a sessile subglobose berry; style persisting at apex. Seed solitary, ± ellipsoid; testa shiny brown with pale narrow lateral scar; endosperm absent; cotyledons plano-convex; radicle basal.

A monotypic genus closely related to *Aningeria* and sharing with it the pellucid-punctate leaf-character. Distinguished however by the subsessile flowers, lack of staminodes and the narrow lateral seed-scar. Both genera are restricted to the African continent.

M. alnifolia (*Baker*) *Pierre*, Not. Bot. Sapot.: 61 (1891); Engl., E.M. 8: 49 (1904); J. H. Hemsl. in K.B. 15: 284 (1961); K.T.S.: 525 (1961); Heine in F.W.T.A., ed. 2, 2: 24 (1963); Fl. Cameroun 2: 126, t. 27 (1964). Type: S. Nigeria, Onitsha, *Barter* 1788 (K, holo.!)

Shrub* to medium tree, height up to 25 m., with fluted trunk and slightly buttressed base. Hairs of young shoots and petioles brown, dense and ± erect. Petioles up to 2 cm. long. Leaves change colour to orange and red before falling; lamina obovate to oblong, 12–22(–36) cm. long, 7–12(–20) cm. wide, apex rounded, sometimes apiculate, base obtuse or cuneate; upper surface with scattered hairs, becoming subglabrous, midrib channelled and with dense short hispid hairs along length; lower surface with uniformly scattered erect hairs, densely arranged on midrib and nerves or sometimes glabrescent with few hairs only on midrib; lateral nerves 15–26 each side, veins prominent, raised and oblique. Bracts sepal-like, densely rusty hairy. Sepals elliptic to ovate, up to 6 mm. long, 4·5 mm. wide, densely rusty hairy externally. Corolla-lobes up to 3·5 mm. long. Filaments up to 5 mm. long. Style 3–7(–9) mm. long. Ripe fruit red and fleshy, up to 2·5 cm. long, with indumentum persisting as ± felted covering until almost mature. Seed up to 1·7 cm. long, slightly flattened; testa dark brown; scar linear, up to 4 mm. wide. Fig. 3.

var. **alnifolia**

Young shoots and petioles with pale to dark brown pubescence of very densely or densely arranged short and ± erect hairs, rarely petioles becoming glabrescent.

* Not yet recorded as such in East Africa.

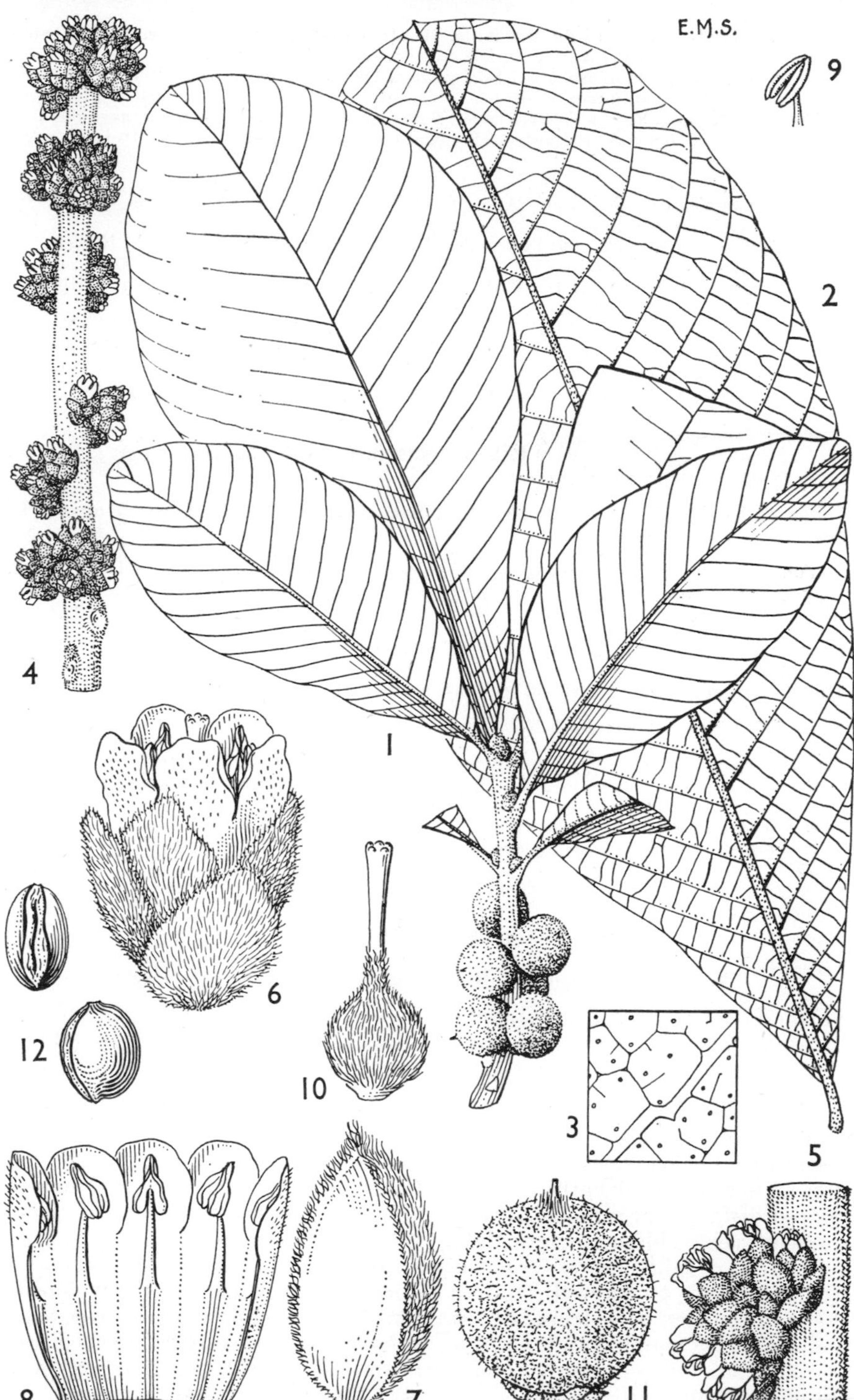

FIG. 3. *MALACANTHA ALNIFOLIA*—**1**, fruiting branchlet, × ½; **2**, leaf, × ½; **3**, lower surface of leaf, × 40; **4**, flowering branch of var. *SACLEUXII*, × 1; **5**, flower-cluster of same, × 2; **6**, flower, × 6; **7**, sepal, × 6; **8**, section of corolla, × 6; **9**, anther, × 6; **10**, ovary, × 6; **11**, fruit, × 1½; **12**, seeds, × 1. 1, 11, 12, from *Semsei* 1954; 2, from *Semsei* 866; 3, from *Drummond & Hemsley* 4035; 4, 5, from *Vaughan* 1967; 6–10, from *Mshatshi* in *Herb. Amani* 3062.

KENYA. Kwale District: Shimba Hills, Mwele Mdogo Forest, 28 Aug. 1953, *Drummond & Hemsley* 4035! & Mrima Hill, 5 Sept. 1957, *Verdcourt* 1881!
TANGANYIKA. E. Usambara Mts., Sigi valley, Dec. 1908 (fr.), *Zimmermann* in *Herb. Amani* G7668!; Handeni District: Misufini, Oct. 1950, *Semsei* 564!; Morogoro District: Mtibwa Forest Reserve [near Turiani], Aug. 1952 (fl. buds), *Semsei* 866!
DISTR. **K**7; **T**3, 6; widespread from Senegal to Sudan Republic and south to Mozambique

SYN. *Chrysophyllum alnifolium* Baker in F.T.A. 3: 499 (1877) [genus queried by original author]
Malacantha heudelotiana Pierre, Not. Bot. Sapot.: 61 (1891); Engl., E.M. 8: 48 (1904); Aubrév., Fl. For. Côte d'Ivoire, ed. 2, 3: 134, t. 299 (1959), sed non sensu Hutch. & Dalz., F.W.T.A. 2: 12 (1931). Type: Gambia, Kombo, *Heudelot* 52 (P, holo.!)
Chrysophyllum ferrugineo-tomentosum Engl. in E.J. 28: 447 (1900). Types: Tanganyika, foothills of Uluguru Mts., *Stuhlmann* 8720 & between Mgeta and Mbakana [Rivers], *Goetze* 338 (both B, syn. †)
Malacantha ferrugineo-tomentosa (Engl.) Engl., E.M. 8: 48, t. 18/A (1904); T.T.C.L: 563 (1949)
M. sp. sensu F. W. Andr., F.P.S. 2: 374 (1952)
Pouteria alnifolia (Baker) Roberty in Bull. I.F.A.N. 15: 1417 (1953) & Pet. Fl. Ouest-Afr.: 80 (1954)

var. **sacleuxii** (*Lecomte*) *J. H. Hemsl.* in K.B. 15: 287 (1961). Type: Zanzibar I., *Sacleux* 1767 (P, holo., K, iso.!)

Young shoots and petioles with a ferrugineous pubescence of very densely arranged, long ± erect hairs, giving a soft furry appearance to these parts, persisting on midrib and lateral nerves of leaf beneath.

ZANZIBAR. Zanzibar I., Haitajwa, 16 Feb. 1930, *Vaughan* 1228! & 18 Sept. 1932 (fl.), *Vaughan* 1967!
DISTR. **Z**; apparently restricted to Zanzibar Island

SYN. *M. sacleuxii* Lecomte in Not. Syst. 4: 62 (1923)

HAB. (of species as a whole). An understory tree in lowland rain-forest, groundwater forest and riverine forest and also in deciduous forest within the coastal belt in East Africa. It appears, however, to be absent from the closed rain-forests of the Congo and West Africa and in the latter region is commonly found in mixed deciduous forest. It is also a frequent colonizing tree in woodland areas in the west; 0–600 m.

4. ANINGERIA

Aubrév. & Pellegr. in Bull. Soc. Bot. Fr. 81: 795 (1935)

[*Pouteria* sensu Baehni in Candollea 7: 418 (1938), pro parte; Meeuse in Bothalia 7: 341 (1960), pro parte; Baehni in Boissiera 11: 48 (1965), pro parte, *non* Aubl.]

Tall trees. Young shoots and petioles pubescent, older branchlets deep purplish or blackish-brown in colour, glabrous. Stipules absent. Leaves petiolate; lamina chartaceous or coriaceous, pellucid punctate; midrib and lateral nerves prominently raised on lower surface, primary lateral nerves looped distally and running parallel to leaf-margin, secondary nerves rarely present, veins oblique or reticulate. Flowers pedicellate, fascicled in axils of current or fallen leaves, or sometimes on short axillary shoots. Sepals 5 (variants with 4 observed), ± free to base. Corolla-tube well-developed; lobes 5, rounded or ± truncate. Stamens 5, inserted on tube, shorter than corolla. Five staminodes present, ± subulate or very rarely flattened and ± petaloid. Style robust, thickened towards apex; stigma 5-papillate. Fruit a subglobose to obovoid or narrowly ellipsoid berry, sometimes rostrate. Seed usually solitary, narrowly to broadly ellipsoid; testa shiny brown, with a pale ± ellipsoid lateral to obliquely lateral scar covering up to half surface area; endosperm absent; cotyledons plano-convex; radicle basal.

A small African genus of about four species, closely related to *Malacantha* (see comment under the latter genus).

All *Aningeria* species are very large trees* with long straight clean trunks. They yield a good quality timber which has a peculiar pungent smell when freshly sawn but this soon disappears with storage.

In the treatment of the species no account is taken of sapling or sucker leaf characters. Juvenile leaves, especially in the case of *A. adolfi-friedericii*, appear to be extremely variable in shape, size and indumentum, often bearing little resemblance to the mature foliage of the crown. The raising of seedlings from authentic seed of known stock trees and preservation of specimens to show the range and development of the resulting leaves are much to be desired. When using the key, leaves only from the uppermost branches of older trees should be used.

Mature leaves usually coriaceous, with lamina opaque and apparently** without system of pellucid dots . . . . 2. *A. adolfi-friedericii*

Mature leaves chartaceous to subcoriaceous, with extensive system of pellucid dots easily visible when viewed against strong light with handlens:

Upper surface of leaves with veins inconspicuous and barely raised; flowers borne on short and densely pubescent axillary shoots . 3. *A. pseudoracemosa*

Upper surface of leaves with vein reticulum raised and easily visible; flowers fascicled in leaf axils . . . . 1. *A. altissima*

1. **A. altissima** (*A. Chev.*) *Aubrév. & Pellegr.* in Bull. Soc. Bot. Fr. 81: 796 (1935); Wimbush, Cat. Kenya Timbers: 27 (1950); I.T.U., ed. 2: 386, fig. 77 (1952); F.P.S. 2: 371 (1952); Verdc. in K.B. 11: 453 (1957); Aubrév., Fl. For. Côte d'Ivoire, ed. 2, 3: 136, t. 300/1–4 (1959); J. H. Hemsl. in K.B. 15: 277 (1961); K.T.S.: 523, fig. 96 (1961); Fl. Gabon 1: 149, t. 26/1–5 (1961); Heine in F.W.T.A., ed. 2, 2: 24 (1963); Fl. Cameroun 2: 134, t. 29/1–4 (1964). Types: Guinée Republic, Kaba valley, *Chevalier* 13129 & subtributary of Mango R., *Chevalier* 13141 (both P, syn., K, isosyn. !)

Tall tree, height up to 50 m., with clean straight cylindrical bole and pale greyish bark, slightly buttressed at base. Young shoots and petioles finely pubescent or puberulous; older branches blackish-brown and glabrous. Petioles up to 1·5 cm. long. Leaf-lamina elliptic to elliptic-obovate or oblong-elliptic, 5–13(–16) cm. long, 3–7 cm. wide, obtuse and emarginate or sometimes shortly and bluntly acuminate, rarely acute, base rounded or abruptly and broadly cuneate, subglabrous except on midrib of lower surface; lateral nerves 14–22 each side, arcuate, ascending. Flowers fragrant, clustered in axils of current leaves, with 2–8 in each axil; pedicels 3–6 mm. long, pubescent. Sepals spreading, elliptic to broadly ovate, 3·5–5·5 mm. long, 2·5–4 mm. wide, pubescent or puberulous outside with very short hairs. Corolla greenish-cream to pale yellow; tube up to 3·5 mm. long; lobes ± ovate to elliptic-oblong, up to 2 mm. long, ciliate. Free part of filaments up to 1·5 mm. long. Staminodes subulate, ± 2 mm. long. Ovary densely pilose; style up to 3·5 mm. long. Fruit red, obovoid to subglobose, up to 2 cm. in diameter, finely pubescent when young, becoming subglabrous at maturity. Seed ± obovoid, up to 1·5 cm. long; testa shiny brown; scar pale and rough, ± elliptic.

* This does not include *A. pseudoracemosa*, details of which are still very inadequate.

** These are present and visible in thin textured sapling and sucker leaves, but obscured by thickness of mature leaves.

UGANDA. W. Nile District: Payida [Paida], Feb. 1934 (fl.), *Eggeling* 1536!; Bunyoro District: Budongo Forest, Feb. 1932 (fl.), *Harris* 53 in *F.D.* 460!; Mengo District: Kajansi Forest, 16 km. on [Kampala–] Entebbe road, Oct. 1937 (fl.), *Chandler* 1994!
KENYA. N. Kavirondo District: Kakamega Forest, June 1934 (fl.), *Dale* in *F.D.* 3257! & 11 Dec. 1956, *Verdcourt* 1684!
TANGANYIKA. Biharamulo District: Nyakanazi, July 1953 (fl.), *Eggeling* 6639!; Mwanza District: Geita, 14 Apr. 1937, *B. D. Burtt* 6510!
DISTR. **U**1–4; **K**5; **T**1; extends from Guinée Republic in the west through the Ivory Coast, Ghana and Cameroun Republic to the Sudan Republic and SW. Ethiopia
HAB. An upper canopy tree of the Uganda lowland rain-forests, it regenerates freely and is a common dominant species, also occurring in riverine forest; 1000–1700 m.

SYN. *Hormogyne altissima* A. Chev. in Mém. Soc. Bot. Fr. 8: 265 (1917) & Veg. Ut. Afr. Trop. Fr. 9: 263 (1917)
Sideroxylon altissimum (A. Chev.) Hutch. & Dalz., F.W.T.A. 2: 12 (1931)
Pouteria giordani Chiov. in Atti R. Accad. Ital. 11: 43 (1940). Type: Ethiopia, Galla Sidamo, near Dembidollo, Uaba Forest, *Giordano* 2455 (FI, holo.!)
P. altissima (A. Chev.) Baehni in Candollea 9: 292 (1942)

NOTE. The flowers of the Uganda material are on the whole slightly larger than those from West Africa, but this may be regarded as normal size variation in a species showing a characteristic East to West African forest distribution. Large resources of the tree, providing a good quality and easily worked timber, occur in Uganda. The timber is marketed under the trade name of Osan or Osen. In Kenya, where the species is known from the Kakamega Forest area only, the timber goes under the name Mukangu and is said to resemble that of Muna, *Aningeria adolfi-friedericii*, very closely.

2. **A. adolfi-friedericii** (*Engl.*) *Robyns & Gilbert* in F.P.N.A. 2: 43 (1947); I.T.U., ed. 2: 385 (1952); K.T.S.: 521 (1961). Types: Rwanda Republic, Bugoie Forest, *Mildbraed* 1447 & 1481 & Congo Republic, Ruwenzori, Butagu valley, *Mildbraed* 2528 (all B, syn. †)

Tall tree, height up to 50 m., with long straight ± fluted bole and buttressed base. Young shoots with dense ferrugineous pubescence; older branches blackish-brown and subglabrous. Petioles 1–2 cm. long, twisted, with ferrugineous or greyish-brown pubescence. Leaf-lamina elliptic to oblong or obovate-elliptic, 4–21·5 cm. long, 2–8·5 cm. wide, pellucid dots visible only in thin-textured sapling or youngest shade leaves, apex acute or rounded, base narrowly to broadly cuneate and sometimes decurrent with petiole, margin not or strongly inrolled*; lower surface with varying density of ferrugineous pubescence, wearing away or stripping and becoming ± glabrous and with hairs on midrib and nerves only; lateral nerves 10–25 each side. Flowers fascicled in current or fallen leaf axils; pedicels 5–10 mm. long, densely pubescent. Sepals 4 or 5, ovate to ± oblong, up to 6 mm. long, 3 mm. wide, pubescent externally with hairs of variable density and size. Corolla cream; tube up to 6·5 mm. long; lobes ± ovate or rounded, up to 2 mm. long. Free part of filaments up to 1·5 mm. long. Staminodes subulate or sometimes expanded and petaloid, up to 1·5 mm. long. Style up to 6·5 mm. long. Fruits greenish, narrowly ellipsoid, up to 4 cm. long, apex with short ± 1 cm. long beak, pubescent or puberulous with short ferrugineous hairs. Seed ovoid to ± narrowly ellipsoid, up to 3 cm. long; testa shiny brown; scar lateral, pale, elliptic. Fig. 4.

NOTE. In general aspect, leaves of this species are easily confused with those of *Chrysophyllum gorungosanum* Engl. (*C. fulvum* S. Moore), both species frequently occurring together in upland rain-forest. The indumentum on the lower surface of the leaves provides a ready means of separation, that of the present species consisting of tangled erect or spreading hairs whereas the hairs of *C. gorungosanum* are appressed and ± unidirectional.

* Mainly in leaves from positions of high exposure and full light intensity. The curved and inrolled leaf seems to be more frequent in the subsp. *keniensis* and *usambarensis* than in the other subspecies.

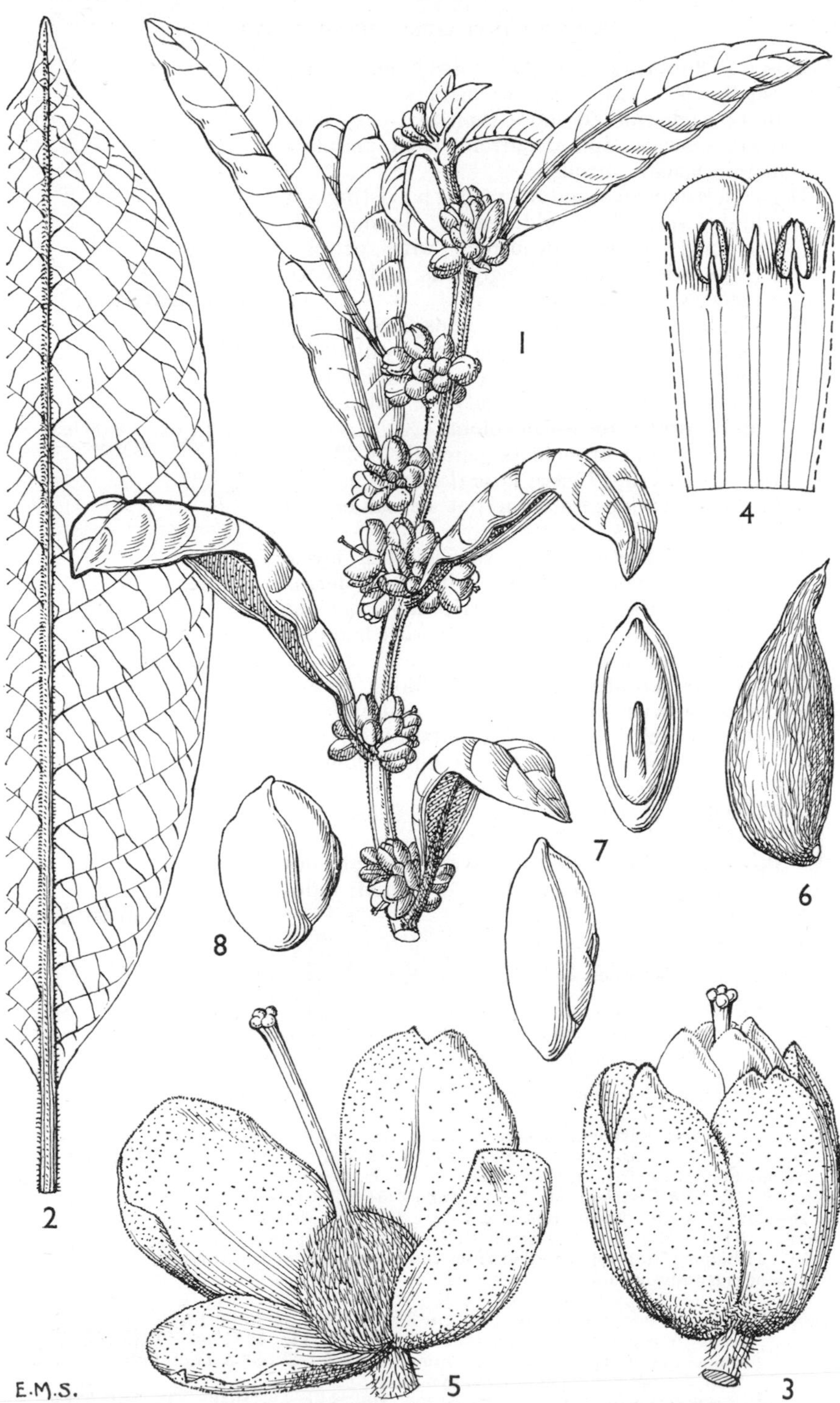

FIG. 4. *ANINGERIA ADOLFI-FRIEDERICII*—**1,** flowering branchlet of subsp. *KENIENSIS*, × 1; **2,** leaf of subsp. *ADOLFI-FRIEDERICII*, × 1; **3,** young flower of subsp. *KENIENSIS*, × 6; **4,** section of corolla, × 6; **5,** flower with corolla removed, × 6; **6,** fruit of subsp. *ADOLFI-FRIEDERICII*, × 1; **7, 8,** seeds, × 1. 1, from *Hockliffe* in *F.D.* 1369; 2, 8, from *St. Clair-Thompson* in *Eggeling* 3956; 3–5, from *Wye* in *F.D.* 929; 6, 7, from *Greenway & Hummel* 7308.

KEY TO INTRASPECIFIC VARIANTS

Leaves of flowering branchlets 4–12 cm. long, usually 5–9 cm.; lateral nerves generally 10–14 each side (if more then leaf-base narrowly cuneate and decurrent with petiole); margin usually inrolled:
 Dense ferrugineous indumentum persisting on lower surface of old leaves; lamina not or only slightly decurrent with petiole; calyx with dense ferrugineous pubescence externally . . . . subsp. **keniensis**
 Ferrugineous indumentum conspicuous on youngest leaves only, older leaves pubescent but lower surface ± green in colour; lamina decurrent with petiole; calyx ± puberulous and greenish-brown in colour . . . subsp. **usambarensis**
Leaves of flowering branchlets generally 8–22 cm. long, usually 9–17 cm.; lateral nerves usually 14–25 each side; margin not or sometimes inrolled:
 Leaf-lamina elliptic to obovate-elliptic, lower surface of younger leaves with dense ferrugineous pubescence, later stripping away and surface becoming ± glabrous with hairs on midrib and nerves only . . . . subsp. **australis**
 Leaf-lamina elliptic to elliptic-oblong, lower surface with dense or very dense ferrugineous or orange-brown pubescence, persisting but sometimes changing colour to greyish-brown in old leaves:
  Pedicels and calyces with very dense pale brown floccose indumentum . . . subsp. **floccosa**
  Pedicels and calyces with dense ferrugineous or greyish-brown close ± felted indumentum . . . . subsp. **adolfi-friedericii**

subsp. **adolfi-friedericii**; J. H. Hemsl. in K.B. 15: 279 (1961)

Petioles (1–)1·5–2·3 cm. long. Leaf-lamina elliptic to elliptic-oblong, mostly 9–21·5 cm. long and 4–7·5 cm. wide, margins not or sometimes inrolled, lower surface with dense ferrugineous pubescence, persisting in older leaves but sometimes becoming greyish-brown in colour; lateral nerves 13–20, usually 14–18 on each side. Pedicels and calyces with dense ferrugineous or greyish-brown ± felted indumentum. Fig. 4/2, 6–8.

UGANDA. Acholi District: Imatong Mts., Apr. 1938 (fr.), *Eggeling* 3554!; Toro District: Ruwenzori, Bwamba Pass, July 1940 (fl.), *Eggeling* 3996!; Mbale District: Elgon, Feb. 1940 (fr.), *St. Clair-Thompson* in *Eggeling* 3956!
KENYA. W. Suk District: Mt. Mtelo, June 1954 (fl.), *Colby* H134!; Kisumu–Londiani District: Tinderet Forest Reserve, 7 July 1949 (fl.), *Maas Geesteranus* 5399!
DISTR. **U**1–3; **K**2, 3, 5; extreme east of Congo Republic, Rwanda Republic and SW. Ethiopia; not yet recorded from the Sudan side of the Imatong Mts., where it almost certainly occurs

SYN. *Sideroxylon adolfi-friedericii* Engl. in Z.A.E.: 519, t. 70 (1913)
Pouteria ferruginea Chiov. in Atti R. Accad. Ital. 11: 42 (1940). Types: Ethiopia, Galla Sidamo, Humbi [? Umbi] Forest, *Giugliarelli* 583 & Dulli Forest, *Giordano* 2447 & *Capuano* 2530 (all FI, syn.!)
P. rufinervis Chiov. in Atti R. Accad. Ital. 11: 44 (1940). Types: Ethiopia, Galla Sidamo, Mugghi [? Moghi] Forest, *Giordano* 2478 & Bosco di Saio, *Giugliarelli* 564 (both FI, syn.!)

[*Malacantha superba* sensu auct. *non* Vermoesen (see T.T.C.L.: 563 (1949)), based on *Pitt-Schenkel* 244 !]
Pouteria adolfi-friedericii (Engl.) Meeuse in Bothalia 7: 341 (1960); Baehni in Boissiera 11: 58 (1965)

subsp. **keniensis** (*R. E. Fries*) *J. H. Hemsl.* in K.B. 15: 279 (1961). Type: Mt. Kenya, S. side, *Fries* 2059 (UPS, holo., K, iso. !)

Petiole 1–1·5(–1·8) cm. long. Leaf-lamina ± elliptic to elliptic-oblong (shape frequently distorted by curved midrib and strongly inrolled leaf-margins), 4–9(–12) cm. long, 2–7 cm. wide, base not or only slightly decurrent with petiole, lower surface densely ferrugineous pubescent; lateral nerves 10–15, usually 12–14 on each side. Calyx with dense ferrugineous pubescence externally. Fig. 4/1, 3–5.

KENYA. Mt. Kenya, S. side, *Hockliffe* in *F.D.* 1369 ! & Ragati [? R. Ruguti], Aug. 1936 (fl.), *Porter* 1021 !; Kiambu District: Escarpment, *Wye* in *F.D.* 929 !
DISTR. **K3**, 4; ? **T2**; not recorded elsewhere

SYN. *Sideroxylon adolfi-friedericii* Engl. subsp. *keniensis* R. E. Fries in N.B.G.B. 9: 332 (1925)
Malacantha sp. near *M. alnifolia* Pierre sensu Wimbush, Cat. Kenya Timbers: 50 (1950)

NOTE. The main features of the subspecies are the small leaves with lengths about twice that of the width and the conspicuously curled and inrolled lamina appearing to be the normal condition for crown foliage. Flowers generally are a little smaller than in subsp. *adolfi-friedericii*, the latter usually showing the maximum dimensions recorded for the species as a whole. Pedicels are also less robust. The record for **T2** is based on a gathering from the Themi R., near Arusha, at ± 1430 m., *Lindeman* 866 !, tentatively suggested to be a lush-growing low-altitudinal form of the present subspecies (see K.B. 15: 282 (1961)).

subsp. **usambarensis** *J. H. Hemsl.* in K.B. 15: 281 (1961). Type: Tanganyika, W. Usambara Mts., Shagai Forest near Sunga, *Drummond & Hemsley* 2592A (K, holo. !, EA, iso. !)

Petiole 5–13 mm. long. Leaf-lamina narrowly to broadly elliptic, 6–13 cm. long, 2·5–4 cm. wide, base narrowly cuneate and decurrent with petiole; lower-surface with ferrugineous indumentum only in youngest leaves, becoming thinly pubescent and greenish in colour in older leaves; lateral nerves 11–16 each side. Calyx greenish-brown, thinly pubescent to puberulous with very short hairs.

TANGANYIKA. W. Usambara Mts., Shagai Forest, May 1953 (fl.), *Procter* 215A ! & Baga Forest, Nov. 1936, *Markham* 987 ! & E. Usambara Mts., Kwamkoro Forest, 21 Jan. 1960, *Msuya* 12 !
DISTR. **T2**, 3; not known elsewhere

NOTE. Mature leaves are not unlike those of subsp. *keniensis* in general form, but differ in the relatively sparse indumentum and decurrent leaf-bases. The present taxon is the least hairy of all the subspecies, and the calyces at anthesis are never more than thinly pubescent.

subsp. **floccosa** *J. H. Hemsl.* in K.B. 15: 280 (1961). Type: Tanganyika, Mbulu District, Oldeani, *Hitchcock* 3 (K, holo. !, EA, iso. !)

Petiole 1–2·6 cm. long. Leaf-lamina elliptic to oblong-elliptic, 11–17 cm. long, 4–6·5 cm. wide, lower surface with very dense ferrugineous or orange-brown pubescence, margin not or slightly inrolled; lateral nerves 15–21 each side. Pedicels and calyx with dense pale brown floccose indumentum.

TANGANYIKA. Mbulu District: Oldeani, above Mr. Major's Farm, 23 Mar. 1950 (fl.), *Hitchcock* 3 !
DISTR. **T2**; not recorded elsewhere

NOTE. The gathering is stated to come from altitudes 3050–3200 m., which are at the extreme summit of the mountain. This is unusually high for a timber tree and the specimen shows none of the epiphytic lichen infestation normally associated with such habitats. It is felt that the information requires confirmation. *Carmichael* in *E.A.H.* 13416 (EA !), from Mt. Hanang in the same district, probably also belongs here, although in the absence of flowers and fruits it is not certainly determinable.

subsp. **australis** *J. H. Hemsl.* in K.B. 15: 282 (1961). Type: Tanganyika, Rungwe Forest Reserve, *Semsei* 1532 (K, holo. !, EA, iso. !)

Petioles 2–2·7 cm. long. Leaf-lamina elliptic to obovate-elliptic, 12–20 cm. long, 5–8·5 cm. wide, margins not or very slightly inrolled; lower surface of younger leaves with ± dense ferrugineous indumentum, later stripping away and surface becoming practically glabrous with hairs on midrib and nerves only; lateral nerves 18–25 each side. Calyces and pedicels with dense brown felted indumentum.

TANGANYIKA. Ufipa District: Mbisi Forest, Nov. 1958, *Napper* 1077!; Rungwe Forest Reserve, Jan. 1954 (fl.), *Semsei* 1532!; Njombe District: Livingstone Mts., June 1951, *Nightingale* 1!

DISTR. **T**4, 7, 8; Zambia/Malawi border (Nyika Plateau)

SYN. [*A. adolfi-friedericii* sensu F. White, F.F.N.R.: 320 (1962), *non* (Engl.) Robyns & Gilbert sensu stricto]

HAB. (of species as a whole). Upland rain-forest, frequently associated with *Podocarpus* spp., rarely in riverine forest; 1430–2500 (?–3200*) m.

NOTE. The species has been obtained in latter years from a steadily increasing number of localities and from information available may be expected in most upland rain-forest sites within the area of the Flora.

The timber of subsp. *keniensis*, widely known under the name Muna in Kenya is a popular joinery wood, while that of the subsp. *adolfi-friedericii* finds a ready sale when available in Uganda. The species would appear to have good potentiality as a montane timber tree throughout East Africa. There are grounds for considering the species in the wide sense as having given rise to a number of divergent segregates under conditions of geographical discontinuity and ecological isolation; these have been recognized as subspecies by the present author and perhaps represent species in the making. Fruits of the subspp. *usambarensis*, *floccosa* and *australis* are unknown; further collecting of all three would be a welcome addition to present knowledge. *Aningeria adolfi-friedericii* in a broad sense occurs in **K**1 (Mathews Range, Mandasion Range, *Kerfoot* 2618), but the sole record is inadequate to determine the subspecies with certainty.

3. **A. pseudoracemosa** *J. H. Hemsl.* in K.B. 15: 283 (1961). Type: Tanganyika, Pangani District, Bushiri, *Faulkner* 729 (K, holo.!, EA, iso.!)

Tree with smooth greyish or greyish-brown bark; height up to 20 m. Terminal buds, young shoots and petioles with dense ferrugineous indumentum; older twigs with deep purplish-brown bark, glabrous. Petioles 5–12 mm. long. Leaf-lamina elliptic to obovate-elliptic, 8–17 cm. long and 4–7·6 cm. wide, apex rounded to subacute, base rounded to broadly cuneate; margin slightly to conspicuously inrolled; lower surface softly pubescent when young, later becoming puberulous to glabrescent, indumentum usually dense and ferrugineous on midrib and nerves; lateral nerves 14–20 each side. Inflorescence raceme-like with flowers fascicled along a short ± erect or deflexed axillary shoot; rhachis up to 8 cm. long, densely ferrugineous pubescent; bracts, pedicels and flower-buds with dense floccose hairs; bracts subulate to narrowly ovate, up to 4 mm. long; pedicels 1–5 mm. long. Outer sepals ± ovate to elliptic, innermost ± oblanceolate, 4·5–6 mm. long, 2–3 mm. wide. Corolla-tube 4·5–5·5 mm. long, ± 3 mm. in diameter; lobes rounded to bluntly obtuse, up to 2 mm. long, ciliate. Free part of filament 1–1·5 mm. long. Staminodes ± subulate, narrowly oblong or ± spathulate, 1–1·5 mm. long. Ovary subconical, up to 2 mm. in diameter; style 5–8 mm. long; stigma 5-papillate. Fruits unknown.

TANGANYIKA. E. Usambara Mts., Kihuhwi, 14 Nov. 1944, *Flamwell* in *Herb. Amani* H87/44! & Monga, 31 May 1945, *Czurn* B22!; Morogoro District: Kimboza Forest Reserve, July 1952 (fl.), *Semsei* 740!

DISTR. **T**3, 6; not known elsewhere

HAB. Lowland rain-forest; 10–1000 m.

NOTE. The possession of the specialized short axillary flower-bearing shoots serves to distinguish this species from all other East African *Sapotaceae*. The condition is foreshadowed in *A. adolfi-friedericii* where a single gathering, *Hockliffe* in *F.D.* 1369, from

* Based on subsp. *floccosa* record.

Mt. Kenya, shows a parallel development of young axillary flower-bearing shoots near the branch apex, but some of these bear typical foliage leaves and differ but slightly from normal axillary branchlets.

5. SIDEROXYLON

L., Sp. Pl.: 192 (1753) & Gen. Pl., ed. 5: 89 (1754); Engl., E.M. 8: 25 (1904)

Trees or shrubs, sometimes spiny. Stipules generally absent. Leaves distinctly petiolate; lamina usually thick and leathery; lateral nerves inconspicuous (at least in Africa), nervation reticulate. Flowers congested in axils of current or fallen leaves, sometimes cauliflorous, pedicellate. Sepals 5 (abnormally 4 or 6), ± free or very shortly connate at the base; lobes ± ovate or rounded. Corolla-lobes 5 (but see above), connate at base; tube well developed, frequently subequal to lobes in length. Stamens 5, as long as or a little longer than corolla. Staminodes present, petaloid or narrower and ligulate, with serrulate or irregularly laciniate margins. Ovary subglobose to conical, pilose, 5-locular; style robust, tapering to simple or ± capitate stigma. Fruit a single-seeded subglobose berry, with persistent style; skin thin with fleshy mesocarp. Seed subglobose or ± ovoid, with thickened stony wall; testa shiny brown; scar suborbicular or elliptic, basal; endosperm present; cotyledons flattened and foliaceous; radicle basal or lateral.

A rather small reasonably well-defined genus in Africa and the surrounding islands, but inclusion of species from the Far East and from the American continent is very much in dispute. Recent opinion seems to favour treating species from the American continent under distinct genera and to restrict the genus *Sideroxylon* to species with a small and ± basal seed-scar. Professor Aubréville, in Adansonia 3: 29–38 (1963), divides the Old World species into several further genera, restricting *Sideroxylon* to the E. coast of Africa and the Mascarene Is.

S. inerme *L.*, Sp. Pl.: 192 (1753); Engl., E.M. 8: 27, t. 7/B (1904); Sim, For. Fl. Cape Col.: 252, t. 95 (1907); Meeuse in Bothalia 7: 323 (1960) & in F.S.A. 26: 33 (1963); Aubrév. in Adansonia 3: 30, t. 4/1 (1963). Type: probably a Cape plant cultivated in Holland, *Herb. Linnaeus* 266.1 (LINN, lecto.!)

Spreading much-branched evergreen shrub or small tree; height up to 15 m.; older branches glabrous; bark grey and fissured. Petiole 0·5–1·5(–2) cm. long, glabrous. Leaf-lamina elliptic, elliptic-obovate to obovate, (3–)4–10(–15) cm. long, (1·5–)2–6(–7·5) cm. wide, apex obtuse to rounded or emarginate, base tapering, narrowly cuneate and decurrent with petiole, glabrous on older leaves; upper surface with slightly raised nervation, lower surface with lateral nerves slightly raised and inconspicuous. Pedicels up to 7 mm. long, puberulous. Sepals broadly ovate, up to 2·5 mm. long, puberulous or ± glabrescent outside. Corolla greenish-white; tube up to 1·5 mm. long; lobes ± ovate, up to 2·5 mm. long. Filaments 1·5–4 mm. long. Staminodes petaloid, ± ovate, but sometimes constricted near base, 1·5–3 mm. long, irregularly serrulate or rarely ± entire. Ovary subglobose; style short, up to 1·5 mm. long. Fruits green at first but finally ripening black, 6–15 mm. in diameter. Seeds solitary, depressed subglobose, 4–9 mm. in diameter, with 1–5 longitudinal ridges and with 1–4 small pits (impressions of aborted ovules) near the larger basal scar; embryo horizontal, with lateral radicle.

DISTR. (of species as a whole). **K4**, 7; **T3**, 6, 8; **Z**; **P**; Aldabra Is., Somali Republic and Mozambique to Cape Province of South Africa

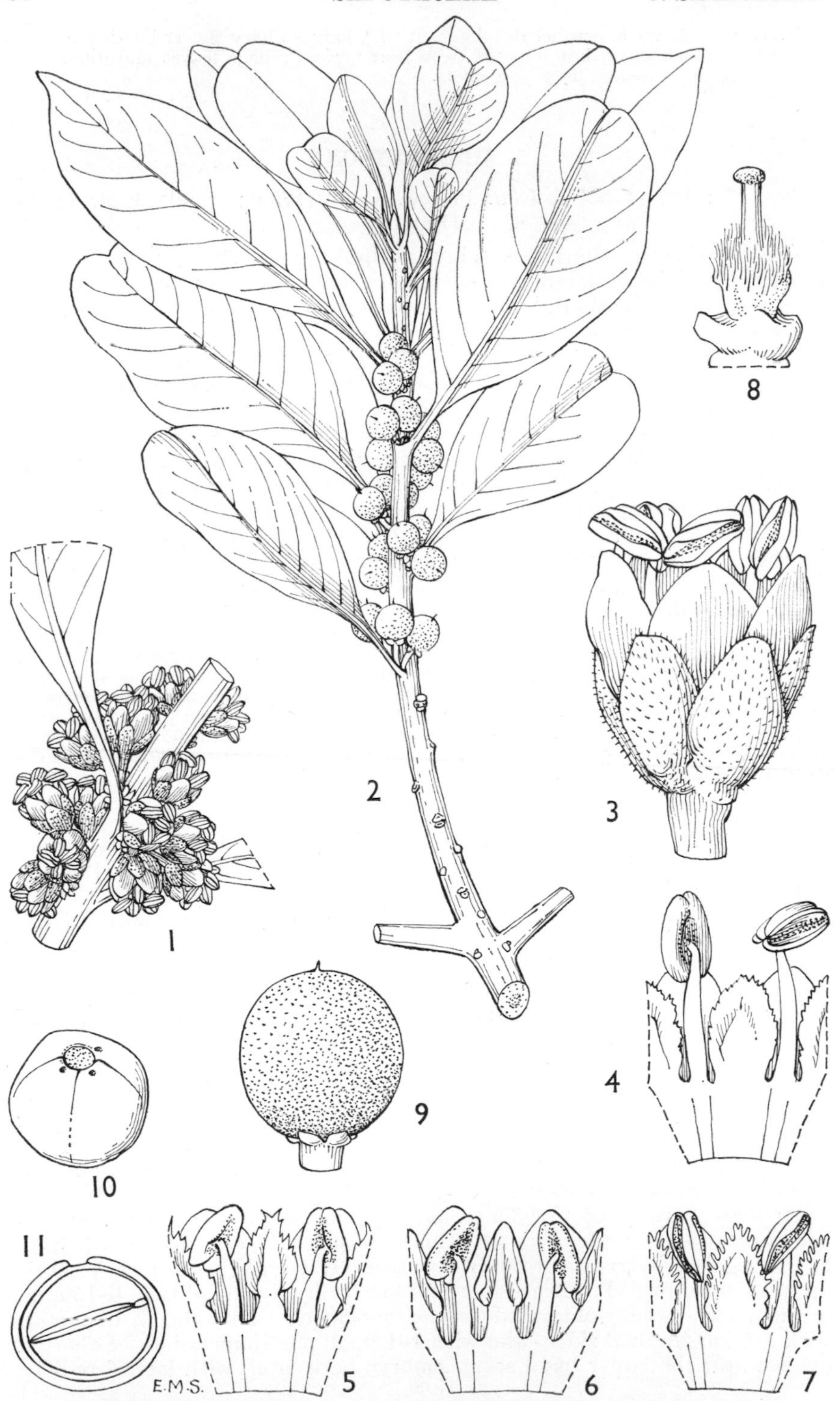

FIG. 5. *SIDEROXYLON INERME* subsp. *DIOSPYROÏDES*—**1**, part of flowering branchlet, × 2; **2**, fruiting branch, × ⅔; **3**, flower, × 10; **4–7**, sections of corollas with stamens and staminodes variously developed, × 10; **8**, ovary and receptacle, × 10; **9**, fruit × 3; **10**, seed in basi-lateral view, × 3; **11**, transverse section of seed, × 3. 1, 3, 4, 8, from *Faulkner* 727; 2, from *Drummond & Hemsley* 3246; 5, from *Gillett* 4516; 6, from *Kassner* 422; 7, from *Boivin*; 9–11, from *Drummond & Hemsley* 3534.

NOTE. The species in the broad sense comprises three subspecies, the following being the only one in East Africa. See discussion in K.B. 20: 472–478 (1966).

subsp. **diospyroïdes** (*Baker*) *J. H. Hemsl.* in K.B. 20: 476 (1966). Type: probably Zanzibar I., *Kirk* 30 (K, holo.!)

Leaf-lamina obovate or rarely elliptic-obovate. Pedicels short, up to 3 mm. long. Filaments 1·5–2 mm. long. Fruit up to 1 cm. in diameter. Seeds 5–7 mm. in diameter, usually with only one ridge developed. Fig. 5.

KENYA. Mombasa, English Point, 26 May 1934 (fl. and yng. fr.), *Napier* in *C.M.* 6411! & in *C.M.* 6272!; Lamu District: Kiunga, 25 Nov. 1946 (fl.), *J. Adamson* 263 in *Bally* 5956! & Takwa Creek, Manda I., 8 Feb. 1956 (fl.), *Greenway & Rawlins* 8874!

TANGANYIKA. Tanga District: 11 km. NE. of Pangani, Kigombe beach, 11 July 1953 (fr.), *Drummond & Hemsley* 3246!; Pangani District: Bushiri, 18 Oct. 1950 (fl.), *Faulkner* 727!; Lindi, about 6·5 km. N. of township, 9 Dec. 1955 (yng. fr.), *Milne-Redhead & Taylor* 7592!

ZANZIBAR. Zanzibar I., Pwani Mchangani, 26 Jan. 1929 (fl.), *Greenway* 1184! & Pwani Kama, 23 Mar. 1962 (fl.), *Faulkner* 3023!; Pemba I., Verani, 18 Feb. 1929, *Greenway* 1462!

DISTR. **K**4, 7; **T**3, 6, 8; **Z**; **P**; extending northwards to Somali Republic (N.) and southwards to Mozambique

HAB. A common component of shrub thickets on the sea-shore near high-water mark and also along the landward fringes of mangrove formations. Less frequently found inland in coastal bushland and woodland. It has been observed inland in saline depressions in thornbush country of Tanganyika near Mkomazi, and collected in Kenya, Kibwezi District, in riverine thickets along the Kiboko R.; 0–900 m.

SYN. *Myrsine querimbensis* Klotzsch in Peters, Reise Mossamb. Bot. 1: 185 (1861). Type: Mozambique, Niassa, Querimba I., *Peters* (B, holo. †)
Sideroxylon diospyroïdes Baker in F.T.A. 3: 502 (1877); Engl., E.M. 8: 27, t. 7/A (1904); T.S.K.: 121 (1936); T.T.C.L.: 568 (1949); K.T.S.: 530 (1961)
S. inerme L. var. *zanzibarensis* Dubard in Ann. Mus. Col. Marseille, sér 2, 10: 87 (1912). Type: Zanzibar I., *Sacleux* 891 (P, holo.)

6. PACHYSTELA

Engl., E.M. 8: 35 (1904); Lecomte in Bull. Mus. Hist. Nat. Paris 25: 192 (1919), excl. sect. *Zeyherella*

Chrysophyllum sect. *Afro-Chrysophyllum* Engl. in E.J. 12: 520 (1890) & in E. & P. Pf. IV. 1: 149 (1891), pro parte minore

Bakeriella Dubard in Not. Syst. 2: 89 (1911), *nom. nud.*, & in Ann. Mus. Col. Marseille, sér. 2, 10: 26 (1912), *nom. illegit.*, pro parte

[*Pouteria* sensu Baehni in Candollea 7: 472 (1938), pro parte; Meeuse in Bothalia 7: 332 (1960), *non* Aubl.]

Shrubs or trees. Leaves mostly terminal. Stipules present, linear-subulate or setose, rigid, persistent. Petioles short. Leaf-lamina coriaceous; primary nerves arcuate-ascending, not closely spaced, raised and conspicuous on lower surface, secondary nerves present or absent; veins oblique or reticulate. Flowers congested in older leaf axils, or cauliflorous, sessile or pedicellate. Sepals 5, imbricate, shortly connate at base or ± free. Corolla-lobes connate at base, spreading. Stamens 5; filaments inserted at throat, subequal to corolla-lobes in length, flexuous. Staminodes present or absent, irregular in number and variable in shape. Ovary ± ovoid, pilose, 5-locular; style long, robust and ± dilated near apex or slender and tapering to apex; stigma with 5-papillae, 5-furrowed or simple. Fruit a subglobose, ovoid or ellipsoid berry, with persistent style; pericarp fleshy. Seed solitary, ovoid or ellipsoid; testa with prominent scar covering at least half of surface area; endosperm absent; cotyledons fleshy, plano-convex; radicle basal.

Tropical African genus of about six species.

Flowers pedicellate:
- Pedicels up to 2 mm. long; sepals ovate to elliptic-oblong, 1·5–3 mm. wide; leaves rarely more than 7·5 cm. wide, tapering to narrowly cuneate base, or rarely base abruptly obtuse . . . . . . 1. *P. brevipes*
- Pedicels up to 6 mm. long; sepals broadly ovate to suborbicular, 4–5 mm. wide; leaves larger, usually more than 7·5 cm. wide, leaf-base abruptly obtuse to subauriculate or rarely ± cuneate . . . . . . 2. *P. msolo*

Flowers sessile . . . . 3. *P. subverticillata*

1. **P. brevipes** (*Baker*) *Engl.*, E.M. 8: 37 (1904); T.S.K.: 120 (1936); T.T.C.L.: 567 (1949); I.T.U., ed. 2: 402 (1952); F.P.S. 2: 375, fig. 140 (1952); Aubrév., Fl. For. Côte d'Ivoire, ed. 2, 3: 148, t. 305/7–12 (1959); K.T.S.: 529 (1961); Fl. Gabon 1: 110 (1961); F.F.N.R.: 322 (1962); Heine in F.W.T.A., ed. 2, 2: 28, fig. 206 (1963); Fl. Cameroun 2: 83, t. 18/1–6 (1964). Types: Zanzibar (I. or coast of mainland opposite), *Kirk* & Malawi, Lake Nyasa, *Kirk* (both K, syn. !)

Small to medium tree with dense crown, height up to 35 m.; bole often pillared and deeply fluted. Young shoots and petioles with dense short appressed hairs; old branches glabrous. Petioles up to 1 cm. long; stipules subulate, up to 1·2 cm. long. Leaf-lamina oblanceolate to obovate, 9–20(–26) cm. long, 3·5–8(–10) cm. wide, apex acuminate, obtuse or emarginate, tapering to narrowly cuneate or rarely abruptly obtuse base; upper surface glabrous, lower surface with greyish pubescence of small appressed hairs on young leaves, old leaves often practically glabrous; primary lateral nerves 8–14 on each side, secondary nerves absent, veins oblique and inconspicuous. Flowers fragrant. Pedicels up to 2 mm. long. Sepals connate at base, ovate to elliptic-oblong, up to 4 mm. long, 3 mm. wide, pubescent externally. Corolla yellowish-green or cream; tube up to 2 mm. long; lobes ± elliptic to narrowly ovate, up to 4·5 mm. long, 2·5 mm. wide. Free part of filament up to 5 mm. long; anthers narrowly obcordate, dehiscence extrorse; small membranous ovate-lanceolate or minute triangular staminodes sometimes present. Ovary subconical, up to 2 mm. long; style up to 5 mm. long. Ripe fruits yellow or orange-yellow, edible with pleasant acid flavour, ellipsoid, up to 2·5 cm. long, sometimes beaked; skin thin; pulp soft and milky. Seed ellipsoid, up to 2 cm. long; testa shiny brown, with large pale ± elliptic lateral scar covering up to two-thirds of surface area. Fig. 6.

UGANDA. Ankole District: Mbarara, Ruizi Falls, June 1938 (fl.), *Eggeling* 3641!; Masaka District: Byante Central Forest Reserve, June 1951 (fl.), *Philip* 473! & Sese Is., Bugala I., 9 June 1932 (fl.), *A. S. Thomas* 168!

KENYA. Embu District: Mwimbi, near Mbogoli, *Rammel* in *F.D.* 1067!; Kwale District: Shimba Hills, Mwele Mdogo Forest, 28 Aug. 1953 (fl.), *Drummond & Hemsley* 4032! & Makadara Forest, Aug. 1929 (fl.), *R. M. Graham* in *F.D.* 2050!

TANGANYIKA. Bukoba District: Nyakato, Apr. 1935 (fr.), *Gillman* 259!; Kilosa District: Kidodi, Oct. 1951 (fl.), *Eggeling* 6301!; Newala District: Chilangala, *Gillman* 1260!

ZANZIBAR. Zanzibar I., 8 km. S. of Chwaka, Mchocha, 8 May 1931 (fl.), *Vaughan* 1935! & Sengenya Dya, 9 Oct. 1951 (fr.), *R. O. Williams* 102!; Pemba I., Ngezi Forest, 11 Dec. 1930 (fr.), *Greenway* 2711!

DISTR. **U**2, 4; **K**4, 5, 7; **T**1, 3, 4, 6–8; **Z**; **P**; widespread in tropical Africa from Portuguese Guinea to the Sudan Republic and south to Angola, Zambia, Malawi, Mozambique and Rhodesia

HAB. Lowland rain-forest and riverine forest, commonly found on river banks and margins of lakes or other such sites with a high permanent water-table; 0–1500 m.

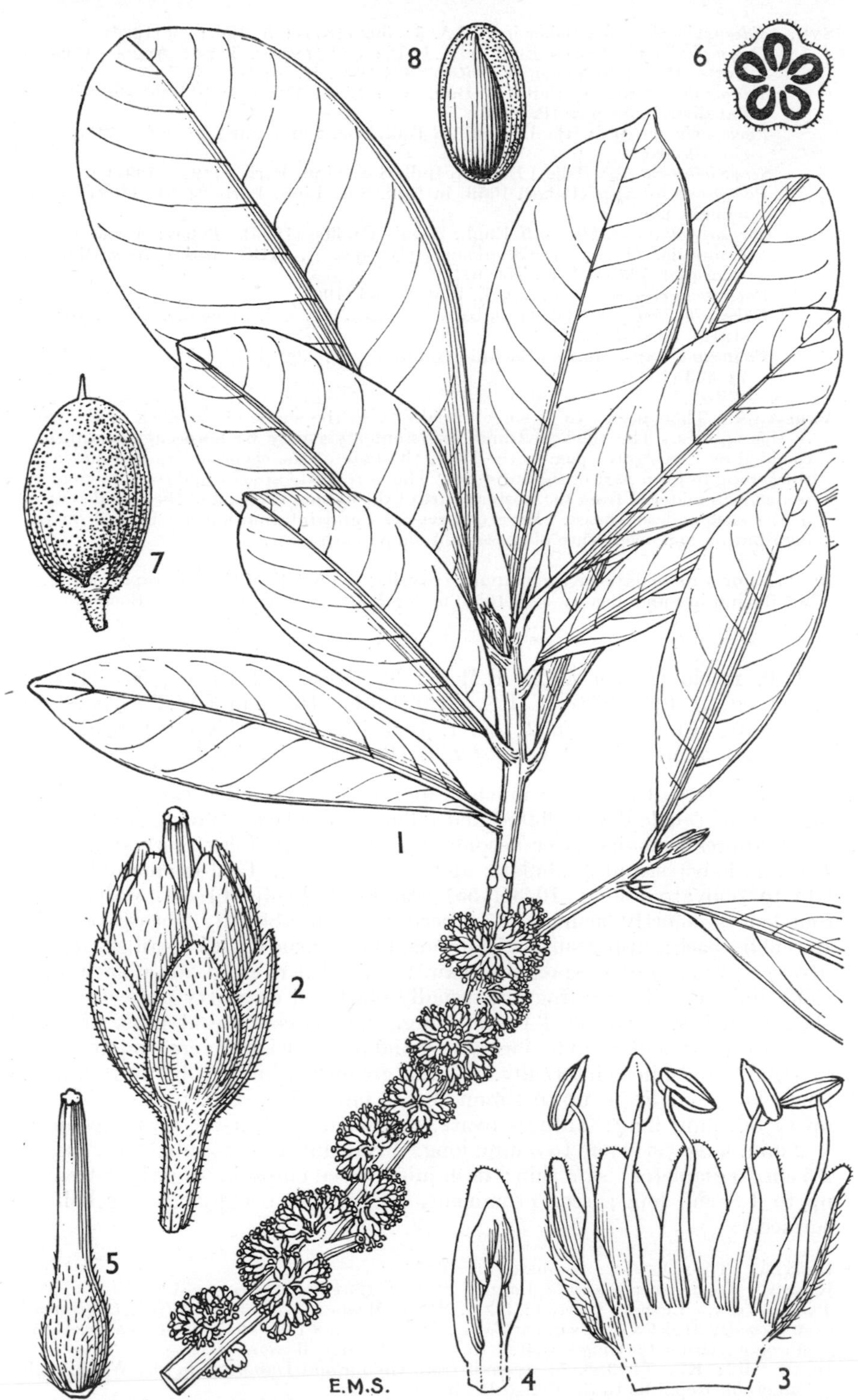

FIG. 6. *PACHYSTELA BREVIPES*—**1,** flowering branch, × ⅔; **2,** young flower, × 6; **3,** corolla dissected to show stamens, × 6; **4,** corolla-lobe and stamen of young flower, × 6; **5,** ovary, × 6; **6,** diagrammatic transverse section of ovary; **7,** fruit, × ½; **8,** seed, × 1½. 1, from *Lyne* 83; 2–6, from *Eggeling* 440 in *F.D.* 774; 7, 8, from *R. O. Williams* 102.

SYN. *Sideroxylon brevipes* Baker in F.T.A. 3: 502 (1877); T.S.K.: 89 (1926)
Chrysophyllum cinereum Engl. in E.J. 12: 522 (1890). Type: Angola, Cuanza Norte, Pungo Andongo, *Welwitsch* 4824 (BM, K, iso. !)
Sideroxylon sacleuxii Baill. in Bull. Soc. Linn. Paris 2: 911 (1891). Type: Zanzibar I., *Sacleux* (P, holo.)
Pachystela sacleuxii (Baill.) Baill. in Bull. Soc. Linn. Paris 2: 946 (1891), *nom. non rite publ.*
Sersalisia brevipes (Baker) Baill. in Bull. Soc. Linn. Paris 2: 947 (1891)
Pachystela brevipes (Baker) Baill. in Bull. Soc. Linn. Paris 2: 947 (1891), *nom. non rite publ.*
Chrysophyllum stuhlmannii Engl., P.O.A. C: 306 (1895). Types: Mozambique, Zambezia, Quelimane, *Stuhlmann* (B, syn. †, HBG, isosyn. !) & Malawi, *Buchanan* 793 (BM, K, isosyn. !)
Pachystela cinerea (Engl.) Engl., E.M. 8: 36 (1904)
Bakeriella brevipes (Baker) Dubard in Ann. Mus. Col. Marseille, sér. 2, 10: 27 (1912), *nom. illegit.*
Pouteria brevipes (Baker) Baehni in Candollea 9: 290 (1942); Meeuse in Bothalia 7: 333 (1960)

VARIATION. There seems to be some variability in the shape of the apex of the fruit in this species. The thickened and persistent style may be borne at the tip of a rounded or ± rostrate apex, in the latter the base of the style undergoing increased thickening to form part of the structure. The ± rostrate apex would seem to be more frequent in material from the western part of the species range, and the rounded apex in the east and southeast. More observations and information are needed on the distribution and occurrence of these fruit-shape variations.

NOTE. For a more extensive synonymy covering the whole of the geographical range, see Baehni in Candollea 9: 290 (1942) under *Pouteria brevipes* (Baker) Baehni.

2. **P. msolo** (*Engl.*) *Engl.*, E.M. 8: 38 (1904); T.T.C.L.: 567 (1949); I.T.U., ed. 2: 403 (1952); K.T.S.: 529 (1961); Heine in F.W.T.A., ed. 2, 2: 28 (1963); Fl. Cameroun 2: 86, t. 18/7–13 (1964). Type: Tanganyika, E. Usambara Mts., Derema, *Holst* 2237 (B, holo. †, K, iso. !)

Medium to tall tree with much-branched and spreading canopy; height up to 50 m.; bole deeply fluted and pillared near base. Young shoots with dense appressed hairs, later becoming glabrous. Petioles short and stout, 4–8 mm. long; stipules subulate, up to 1·5 cm. long. Leaf-lamina oblanceolate to obovate-oblong, 10–35(–55) cm. long, 4–14(–16) cm. wide, apex rounded or shortly acuminate, tapering to an abruptly obtuse or subauriculate base; upper surface glabrous, lower surface with small greyish or silvery ± appressed or spreading hairs; lateral nerves 10–20 on each side, veins oblique. Flowers fragrant, usually clustered on warty projections on older branches. Pedicels 4–6 mm. long. Sepals connate at base, broadly ovate to suborbicular, up to 5 mm. long and 5 mm. wide, cinereous pubescent externally and pilose internally. Corolla greenish-white; tube up to 2 mm. long; lobes ± elliptic, up to 4 mm. long, 3 mm. wide. Free part of filament up to 5·5 mm. long; anthers ovate, dehiscence extrorse. Ovary conical, ± 2 mm. long; style up to 5 mm. long. Fruit dull yellow, subglobose, up to 2·5 cm. in diameter; skin thin; flesh juicy. Seed ellipsoid, slightly flattened, up to 1·8 cm. long; scar prominent, lateral and occupying over half of surface.

UGANDA. Toro District: Semliki, *Dawe* 647 !
KENYA. Tana River District: Baumo, 21 Mar. 1934 (fl.), *Sampson* 51 !
TANGANYIKA. Lushoto District: 5 km. NE. of Mashewa, 7 July 1953 (fr.), *Drummond & Hemsley* 3195 ! & E. Usambara Mts., Amani area, 8 Dec. 1936 (fl.), *Greenway* 4785; Morogoro District: Uluguru Mts., 16 Feb. 1932 (fr.), *Wallace* 495 !
DISTR. **U**2; **K**7; **T**1, 3, 6, 7; extends from Ghana and Dahomey in the W. through Cameroun Republic to the Congo Republic
HAB. Lowland rain-forest, extending into the lower fringes of upland rain-forest, and in riverine forest; 80–1400 m.

SYN. *Chrysophyllum msolo* Engl., Uber Glied Veg. Usambara, in Abhandl. Königl. Akad. Wiss. Berlin 1894: 44, 52 (1894), *nomen subnudum*, & P.O.A. C: 306, t. 37 (1895)
Pachystela ulugurensis Engl. in V.E. 1(1): 360 (1910); T.T.C.L.: 567 (1949), *nomen subnudum*
Pouteria msolo (Engl.) Meeuse in Bothalia 7: 341 (1960)
Amorphospermum msolo (Engl.) Baehni in Boissiera 11: 103 (1965)

3. **P. subverticillata** *E. A. Bruce* in K.B. 1936: 476 (1936); K.T.S.: 529 (1961). Type: Kenya, Lamu District, Utwani Forest, *Mohammed Abdullah* in *F.D.* 3344 (K, holo.!, EA, iso.!)

Shrub or small tree, height up to 8 m.; growth-increase by repeated subapical branching; young shoots and petioles densely pubescent with ± stiffly erect hairs. Leaves ± whorled at ends of long and short shoots. Petioles up to 8 mm. long; stipules setose, up to 1 cm. long. Leaf-lamina obovate to oblanceolate, 5–13 cm. long, 2–6 cm. wide, apex obtuse to emarginate, narrowly cuneate; upper surface glabrous, lower surface pubescent on midrib and nerves of young leaves, later becoming practically glabrous; lateral nerves 7–12 on each side, venation reticulate. Flowers sessile. Sepals closely clasping corolla-tube, ± free to base, broadly ovate, up to 3 mm. long, 2·5 mm. wide, rusty pilose outside. Corolla-tube urceolate, up to 4 mm. long; lobes spreading, ± ovate, up to 5 mm. long and 3 mm. wide. Free part of filament up to 7 mm. long; anthers narrowly obcordate. Staminodes absent. Ovary subglobose, up to 2 mm. long, densely pilose; style long and slender, up to 1·2 cm. long, tapering to simple stigma. Fruits unknown.

KENYA. Kwale District: Gazi, Gogoni Forest, Dec. 1936 (fl.), *Dale* in *F.D.* 3583!; Lamu District: Witu, Gongoni Forest, Oct. 1937 (fl.), *Dale* 1138 in *C.M.* 11481! & Boni Forest, Maungi Pool, 26 Oct. 1957 (fl.), *Greenway & Rawlins* 9431!
DISTR. K7; not known outside this area
HAB. Shrub and lowermost tree layers in lowland rain-forest; 0–100 m.

SYN. *Pseudoboivinella subverticillata* (E. A. Bruce) Aubrév. & Pellegr. (as "*verticillata*") in Not. Syst. 16: 260 (1960)

NOTE. The systematic position of this distinctive species is not entirely settled and a reassessment may be needed when more material becomes available. In general aspect, namely in the method of branching, the indumentum on young stems and petioles, the form of stipules and pattern of leaf nervation, the plant is strongly suggestive of a *Vincentella* species. This is, however, not borne out in flower structure. The plant is maintained for the present as a *Pachystela* species, agreeing reasonably well with flower characters of the latter genus, especially in the long sinuous filaments and general absence of staminodes. Sessile flowers and spiral arrangement of dissimilar sepals are not found in other members of either genus and parallel the distinction between *Malacantha alnifolia* (Baker) Pierre and *Aningeria* species. Fruits are still to be collected and described and may prove to be an important factor in consideration of affinity. The possibility of the species being conspecific with the earlier described but little known *Synsepalum ulugurense* (Engl.) Engl. should not be overlooked (see page 74).

The species is known only from the Kenya coastal forest areas. In the Utwani and Gongoni Forests of the Witu area it is known as Mchambi, while in the Gogoni Forest, Gazi, the vernacular name is Msamvia wa mwitu. Collection of fruiting material is much to be desired.

7. VINCENTELLA

Pierre, Not. Bot. Sapot.: 37 (1891)

[*Sideroxylon* sensu Baker in F.T.A. 3: 502 (1877), pro parte, *non* L.]
Sideroxylon sect. *Bakerisideroxylon* Engl. in E.J. 12: 518 (1890) & in E. & P. Pf. IV. 1: 144 (1890)
[*Sersalisia* sensu Baill., Hist. Pl. 11: 279 (1892), pro parte, *non* R. Br.]
Bakerisideroxylon (Engl.) Engl., E.M. 8: 33 (1904); F.W.T.A. 2: 12 (1931)
Bakeriella Dubard in Ann. Mus. Col. Marseille, sér. 2, 10: 26 (1912), *nom. illegit.*, pro parte

[*Pouteria* sensu Baehni in Candollea 7: 421, 497 (1938), pro parte, *non* Aubl.]

Shrubs or trees. Stipules present, subulate and flattened to filiform, persistent or caducous. Leaves petiolate; lamina with lateral nerves ascending, raised and conspicuous on lower surface, secondary nerves absent, veins reticulate or oblique. Flowers congested in axils of current or fallen leaves, borne on long slender pedicels. Sepals 5, free to base, much smaller than petals, pilose externally especially at thickened apex. Corolla-lobes 5, strongly reflexed in old flowers; tube very short or nil. Stamens 5; filaments slender, ± equal to length of corolla; anthers ovate-cordate, apiculate. Staminodes well-developed, subulate, margin entire or denticulate. Ovary ovoid or subglobose, very densely pilose with long straight hairs, 5-locular; style slender. Fruit an ellipsoid berry, glabrous or puberulous. Seed solitary, ± ellipsoid; testa thin and brittle; scar lateral; endosperm scanty or absent; cotyledons fleshy, plano-convex; radicle basal.

Genus of about six species, confined to tropical and subtropical Africa, mainly in West Africa and Congo Republic.

V. passargei (*Engl.*) *Aubrév.*, Fl. For. Soud.-Guin.: 427, t. 93/1 (1950); F.F.N.R.: 322 (1962); Heine in F.W.T.A., ed. 2, 2: 23 (1963); Fl. Cameroun 2: 100, t. 21/1–5 (1964). Type: Cameroun Republic, Adamawa Massif, *Passarge* 147 (B, holo. †)

Much branched shrub or small tree, sometimes with straggling habit; height up to 8(–15) m. Young branches and petioles with dense ferruginous pubescence; old branches glabrous. Petiole up to 1 cm. long. Stipules caducous, narrowly subulate, up to 1 cm. long, pilose. Leaf-lamina oblanceolate to obovate, up to 12(–19·5) cm. long and 4·5(–6·8) cm. wide, apex shortly acuminate or rounded to emarginate, narrowly to broadly cuneate; upper surface dark green, glabrous, lower surface pale green, sometimes hairy along midrib, otherwise subglabrous; lateral nerves 7–14 on each side; veins reticulate. Pedicels 5–10 mm. long, pilose. Sepals free to base, broadly ovate, up to 2 mm. long. Corolla greenish-white; tube very short; lobes oblong-elliptic, up to 3·5 mm. long and 1·5 mm. wide, strongly reflexed in old flowers. Filaments up to 3·5 mm. long. Staminodes narrowly subulate to filiform, up to 3 mm. long, entire or denticulate with ± flattened base (see fig. 7/5). Ovary subglobose, up to 2 mm. long; style up to 3 mm. long. Mature fruit a yellow or orange-yellow berry, up to 1·5 cm. long, ± glabrous or puberulous; withered flower parts persistent at base; pulp soft and edible. Seed up to 12 mm. long, 8 mm. wide; testa shiny pale greyish-brown with paler ± elliptic lateral scar. Fig. 7.

TANGANYIKA. Buha District: 48 km. S. of Kibondo, Mukugwa R., July 1951 (fl.), *Eggeling* 6211!; Rungwe District: Masukulu Forest, 18 Feb. 1913 (fl.), *Stolz* 1889!; Songea District: R. Luhira about 7 km. N. of Songea, 16 Feb. 1956 (fl.), *Milne-Redhead & Taylor* 8813!

DISTR. T4, 6–8; extends from Guinée Republic eastwards to the western and southern fringes of Tanganyika and southwards to Mozambique, Malawi, Zambia and Angola

HAB. Riverine forest and streamside shrub thickets; 900–1400 m.

SYN. *Bakerisideroxylon passargei* Engl., E.M. 8: 35, t. 11/A (1904)
B. sapinii De Wild. in Rev. Zool. Afr. 7, Suppl. Bot.: 16 (1919). Type: Congo Republic, Katanga, Dilolo, *Sapin* (BR, holo., K, iso.!)
Pouteria tridentata Baehni in Candollea 9: 386 (1942). Type: Tanganyika, Rungwe District, Kyimbila, *Stolz* 1889 (G, holo., BM, K, iso.!)
P. ligulata Baehni in Candollea 9: 386 (1942). Type: Guinée Republic, Kollangui, *Chevalier* 12218 (G, holo., K, iso.!)
Bakerisideroxylon stolzii Mildbr., manuscript name on B, BM and K sheets of *Stolz* 1889

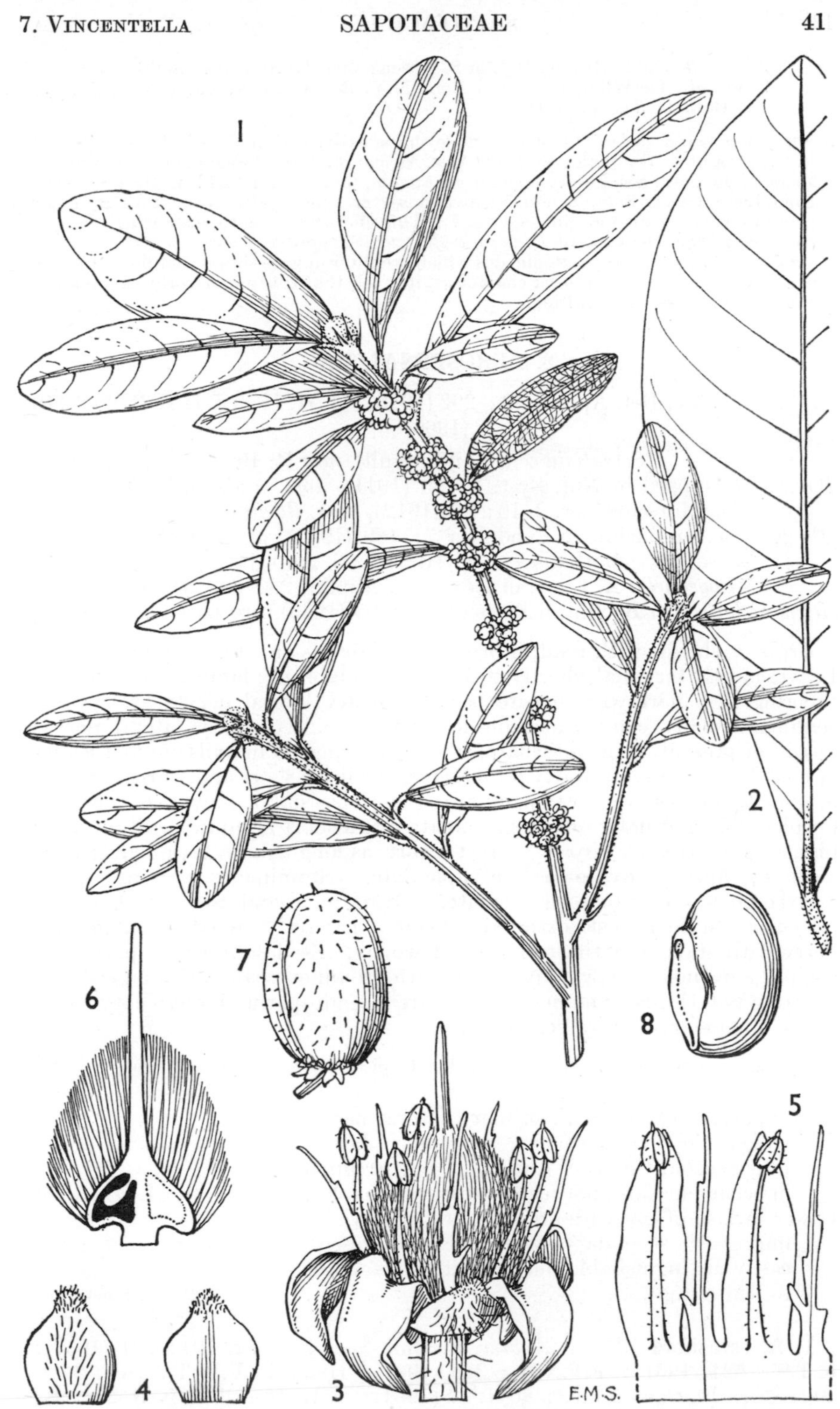

FIG. 7. *VINCENTELLA PASSARGEI*—**1,** flowering branch, × $\frac{2}{3}$; **2,** leaf, × $\frac{2}{3}$; **3,** flower, × 12; **4,** sepals from both aspects, × 12; **5,** section of corolla, × 12; **6,** section of ovary, × 12; **7,** fruit, × 2; **8,** seed, × 2. 1–6, from *Eggeling* 6211; 7, 8, from *Lely* P24.

Vincentella stolzii Hutch., Bot. in S. Africa: 506 (1946), *nomen nudum*
V. sapinii (De Wild.) Brenan in Mem. N.Y. Bot. Gard. 8: 498 (1954); Meeuse in Bothalia 7: 342 (1960)

NOTE. A range in variability of staminode-shape exists within the species as accepted in the present work. The flowers of West African material have a simple ± filiform shape as does also a single specimen from Angola, *Gossweiler* 11767 ! In the specimens from East Africa, Zambia and Malawi, however, staminodes occur with irregular projections along the margins (see fig. 7). This character is not constant and flowers from *Stolz* 1889, the type of *Pouteria tridentata* (the epithet referring to the marginal projections), possess some staminodes which are filiform with simple margins. In view of the good correlation of other characters, it is felt that this minor variation does not warrant taxonomic recognition.

8. AFROSERSALISIA

A. Chev. in Rev. Bot. Appliq. 23 : 292 (1943) ; J. H. Hemsl. in K.B. 20 : 478 (1966)

[*Sersalisia* sensu auct. mult., *non* R. Br.]
Bakeriella Dubard in Not. Syst. 2: 89 (1911), *nom. nud.*, & in Ann. Mus. Col. Marseille, sér. 2, 10: 26 (1912), *nom. illegit.*, pro parte
[*Pouteria* sensu Baehni in Candollea 7: 489 (1938), pro parte; Meeuse in Bothalia 7: 341 (1960), *non* Aubl.]
Rogeonella A. Chev. in Rev. Bot. Appliq. 23: 293 (1943)
[*Richardella* sensu Baehni in Boissiera 11: 95 (1965), pro parte, *non* Pierre]

Trees or shrubs with reddish-brown subglabrous shoots. Stipules absent. Leaves mostly terminal, shortly petiolate and glabrous; lamina ± coriaceous, tapering to a narrowly cuneate base; primary lateral nerves arcuate ascending, usually conspicuous on lower surface, secondary lateral nerves absent or present, veins reticulate. Flowers congested in axils of older leaves or cauliflorous on small warty projections. Pedicels short and sturdy. Sepals 5, connate into a short ± campanulate tube; lobes small, ± ovate or rounded. Corolla-lobes 5, usually reflexing, connate into a short basal tube. Stamens 5; filaments inserted at throat, ± erect, short, as long as or a little longer than anthers; anthers ovate-cordate, apiculate. Staminodes present, small, ± triangular to oblong, denticulate. Ovary ± ovoid to conical, pilose, 5-locular; style robust, exserted. Fruit a narrowly ovoid to subglobose berry with stalk sometimes thick and woody. Seed solitary, ellipsoid and slightly flattened; testa shiny brown with prominent pale rough lateral scar covering two-thirds or more of surface area; endosperm absent; cotyledons fleshy, plano-convex; radicle basal.

A genus with about five species confined to tropical and subtropical Africa.

Lower surface of leaf with primary lateral nerves raised and conspicuous, secondary lateral nerves usually absent, if present very irregular in occurrence and not readily distinguishable . . . 1. *A. cerasifera*
Lower surface of leaf with primary lateral nerves ± impressed, secondary lateral nerves present, easily distinguishable and sometimes extending to leaf-margin . . . . 2. *A. kassneri*

1. **A. cerasifera** (*Welw.*) *Aubrév.* in Bull. Soc. Bot. Fr. 104: 281 (1957); K.T.S.: 521 (1961); F.F.N.R.: 320 (1962); Heine in F.W.T.A., ed. 2, 2: 30 (1963); Fl. Cameroun 2: 80, t. 17 (1964); J. H. Hemsl. in K.B. 20: 482 (1966). Types: Angola, Cuanza Norte, Pungo Andongo, *Welwitsch* 4821 & 4822 (both BM, isosyn. !)

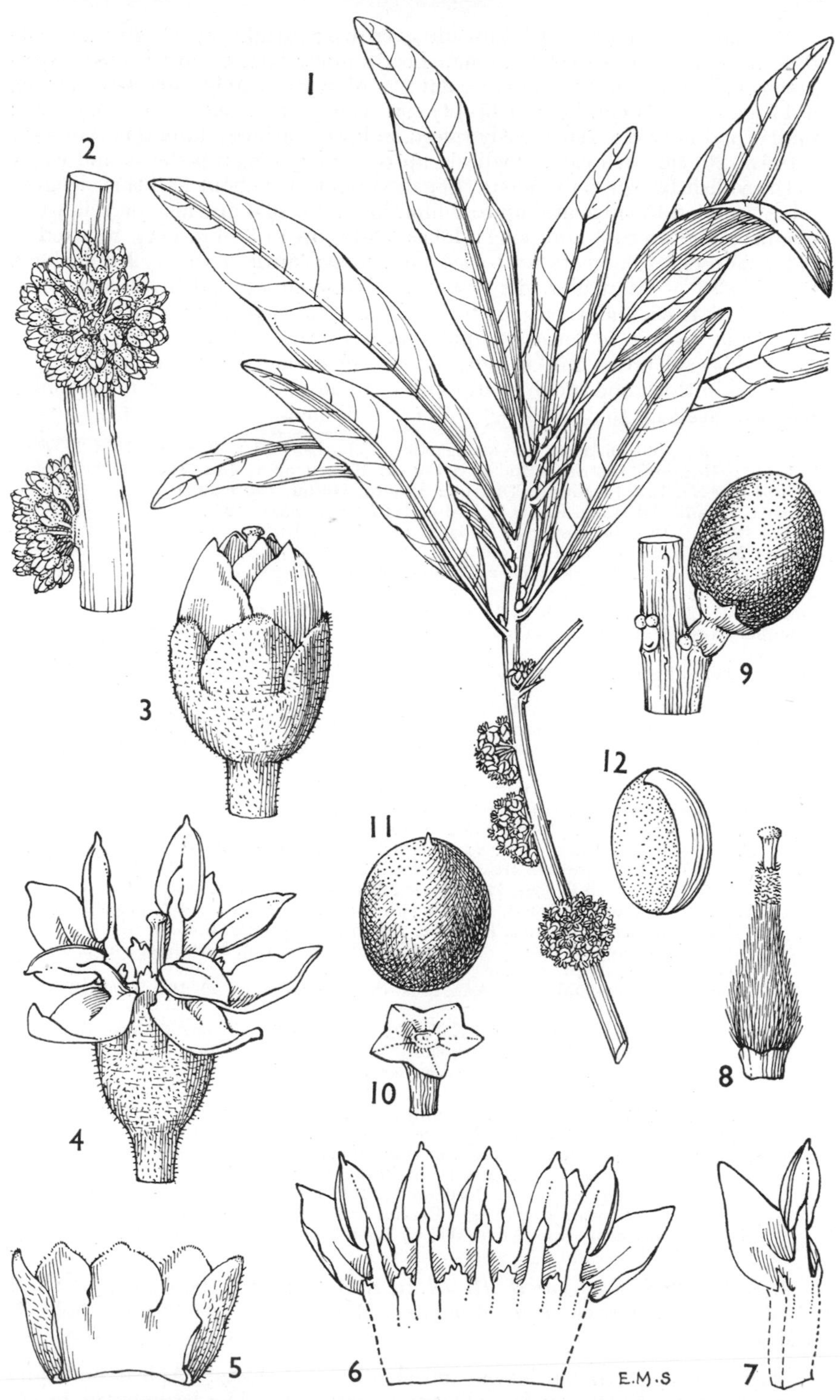

FIG. 8. *AFROSERSALISIA CERASIFERA*—**1,** flowering branchlet, × ½; **2,** clusters of young flowers, × 1; **3,** young flower, × 6; **4,** flower, × 6; **5,** dissected calyx, × 6; **6,** corolla dissected to show stamens and small staminodes, × 6; **7,** corolla segment and stamen, × 6; **8,** ovary, × 6; **9,** fruit, attached, × 1; **10,** fruiting pedicel and cupular calyx, × 1; **11,** fruit, × 1; **12,** seed, × 1. 1, 3–8, from *Semsei* 1406; 2, from *Semsei* 885; 9, from *Greenway* 1004; 10–12, from *Drummond & Hemsley* 3154.

Medium to tall tree with spreading crown; height up to 40 m.; bole fluted. Petioles up to 1·2 cm. long. Leaf-lamina dark green with yellowish-green midrib and nerves, oblanceolate to obovate, rarely narrowly elliptic, 6–17 (rarely –34) cm. long, 5–6(–14) cm. wide, coriaceous, obtuse, cuneate; midrib and nerves conspicuously raised on lower surface; lateral nerves 9–14 (rarely –20) on each side. Pedicels up to 7 mm. long; pedicels and calyx both reddish-brown, puberulous. Sepals connate for about half their length; lobes rounded to subacute, up to 2 mm. long. Corolla greenish or yellowish-cream; tube up to 3 mm. long; lobes ovate, up to 3 mm. long, reflexed in older flowers. Filaments up to 1·5 mm. long; anthers up to 2 mm. long. Staminodes ± triangular, denticulate. Ovary subconical, to 2 mm. long. Fruit-stalk thick and woody, up to 1 cm. long; calyx persistent and becoming thick and woody to form a cupule-like structure. Fruit a red ovoid to globose berry, up to 2·5 cm. long, 2 cm. in diameter; style persisting at apex; skin thick and tough; pulp soft and milky, edible with tart refreshing flavour. Seed up to 2 cm. long, 1·5 cm. in diameter. Fig. 8, p. 43.

UGANDA. W. Nile District: Payida [Paida], Feb. 1934 (fl.), *Eggeling* 1511 in *F.D.* 1441!; Bunyoro District: Budongo Forest, Kisaru area, Jan. 1934 (fr.), *Eggeling* 1481 in *F.D.* 1422!; Masaka District: Buddu, Dumu, *Dawe* 293!

KENYA. Nandi District: Kaimosi, 8 Mar. 1927 (fr.), *Shantz* 125!

TANGANYIKA. Bukoba District: Rubare, July 1951 (fl.), *Eggeling* 6239!; W. Usambara Mts., Kwamshemshi–Sakare road, 4 July 1953 (fr.), *Drummond & Hemsley* 3154!; Morogoro District: near Turiani, Mtibwa Forest Reserve, Aug. 1952 (fl.), *Semsei* 885!

DISTR. **U**1, 2, 4; **K**3; **T**1, 3, 6, 7; extends from Guinée Republic in the west to the Sudan Republic in the north and southwards to Mozambique, Malawi, Zambia and Angola

HAB. Lowland rain-forest, groundwater forest and riverine forest; 300–1500 m.

SYN. *Sapota cerasifera* Welw., Apont.: 585 (1858)
Chrysophyllum disaco Hiern, Cat. Afr. Pl. Welw. 3: 642 (1898). Types: Angola, Cuanza Norte, Golungo Alto, Queta, *Welwitsch* 4812 (BM, K, isosyn.!) & 4820 (BM, isosyn.!)
C. cerasiferum (Welw.) Hiern, Cat. Afr. Pl. Welw. 3: 643 (1898)
Sersalisia disaco (Hiern) Engl., E.M. 8: 30 (1904)
S. cerasifera (Welw.) Engl., E.M. 8: 30 (1904)
S. usambarensis Engl., E.M. 8: 31 (1904); T.T.C.L.: 567 (1949). Types: Tanganyika, E. Usambara Mts., *Scheffler* 172 (B, syn. †, BM, K, isosyn.!) & *Engler* 795 & 799 (both B, syn. †) & *Busse* 2218 (B, syn. †, EA, isosyn.!)
S. edulis S. Moore in J.B. 44: 86 (1906); I.T.U., ed. 2: 403 (1952). Type: Uganda, Mengo District, Entebbe, *Bagshawe* 774 (BM, holo.!)
Bakeriella cerasifera (Welw.) Dubard in Ann. Mus. Col. Marseille, sér. 2, 10: 27 (1912), *nom. illegit.*
B. disaco (Hiern) Dubard in Ann. Mus. Col. Marseille, sér. 2, 10: 27 (1912), *nom. illegit.*
Sersalisia chevalieri Engl. in E.J. 49: 385 (1913). Type: Guinée Republic, *Chevalier* 20168 (B, holo. †, P, iso.)
Pouteria chevalieri (Engl.) Baehni in Candollea 9: 320 (1942)
Rogeonella chevalieri (Engl.) A. Chev. in Rev. Bot. Appliq. 23: 294 (1943)
Afrosersalisia chevalieri (Engl.) Aubrév., Fl. For. Soud.-Guin.: 427 (1950)
A. disaco (Hiern) Aubrév. in Bull. Soc. Bot. Fr. 104: 281 (1957)
A. usambarensis (Engl.) Aubrév. in Bull. Soc. Bot. Fr. 104: 281 (1957)
Pouteria cerasifera (Welw.) Meeuse in Bothalia 7: 341 (1960)
P. disaco (Hiern) Meeuse in Bothalia 7: 341 (1960)
Gymnoluma usambarensis (Engl.) Baehni in Boissiera 11: 101 (1965)
Amorphospermum cerasiferum (Welw.) Baehni in Boissiera 11: 103 (1965)

2. **A. kassneri** (*Engl.*) *J. H. Hemsl.* in K.B. 20: 483 (1966); K.T.S.: 521 (1961). Type: Kenya, near Mombasa, Makoni, *Kassner* 398 (B, holo. †, BM, K, iso.!)

Shrub with ± terminal leaves and slender branchlets. Apical buds and very young shoots with appressed ferrugineous hairs, older branches glabrous. Petioles up to 5 mm. long. Leaf-lamina oblanceolate, up to 13 cm. long,

4 cm. wide, apex shortly and obtusely acuminate to obtuse, tapering to a long, very narrowly cuneate and decurrent base; primary nerves impressed, 9–13 on each side, secondary nerves present, venation reticulate. Pedicels up to 1·5 mm. long; pedicel and calyx both puberulous. Calyx-lobes very small, broadly ovate, up to 1·5 mm. long. Corolla-tube up to 1 mm. long; lobes ovate, up to 1·5 mm. long. Filaments very short. Staminodes very small, ± oblong, dentate at apex. Ovary ovoid, ± 1 mm. long, pilose, tapering to a short style. Fruit not known.

KENYA. Mombasa District: Makoni, 20 Mar. 1902 (fl. buds & fl.), *Kassner* 398!
DISTR. **K**7; ? **T**8; not known elsewhere
HAB. Probably in lowland rain-forest and in riverine forest; less than 300 m.

SYN. *Sersalisia kassneri* Engl., E.M. 8: 31 (1904) [genus queried by original author]
Pouteria kassneri (Engl.) Baehni in Candollea 9: 280 (1942) [genus again queried by this author]

NOTE. The species is known from the type-gathering only. One further specimen, *Busse* 1103! from the Mpatila Plateau in the Makondo area in SE. Tanganyika, has been tentatively assigned to this species by the present author, but as the specimen consists of leaves and very old flowers only, the determination cannot be final. It would seem that the species might occur in other suitable places along the Kenya and Tanganyika coasts and collection of more material (the fruits are as yet unknown) is much to be desired.

9. INHAMBANELLA

(Engl.) Dubard in Ann. Mus. Col. Marseille, sér. 3, 3: 42 (1915), excl. sp. *I. natalensis* (Schinz) Dubard; Aubrév. in Adansonia 1: 6 (1961)

Mimusops L. sect. *Inhambanella* Engl., E.M. 8: 80 (1904)

Small to medium-sized tree. Stipules caducous. Petioles long; lamina glabrous, venation reticulate; primary lateral nerves widely spaced, arcuate, ascending, secondary lateral nerves present. Flowers mostly clustered in axils of current leaves, distinctly pedicellate. Sepals 5–6, free, spirally arranged, broadly ovate, closely imbricate. Corolla exserted; lobes 5, each with a dorsal pair of appendages. Stamens 5, epipetalous; staminodes present, alternating with stamens, petaloid. Ovary conoidal, pilose, tapering into a short thick style. Fruits broadly obovoid to subglobose, firm-walled but fleshy and very milky. Seed solitary, ± oblong to ellipsoid; scar lateral, oblong, extending almost whole length; cotyledons plano-convex; endosperm absent; radicle basal.

A monotypic genus confined to the East African coastal region. It is closely related to *Lecomtedoxa* (Engl.) Dubard,* which comprises several species restricted to the coasts of Gabon and the Cameroun Republic. They differ primarily by the dry coriaceous basally attenuate fruit, which is ultimately dehiscent.

I. henriquesii (*Engl. & Warb.*) *Dubard* in Ann. Mus. Col. Marseille, sér. 3, 3: 43 (1915). Type: Mozambique, Sul do Save, Inhambane, *Ferreira* (COI, holo.!)

Medium-sized tree with copious latex, height up to 25 m.; bark grey. Young shoots with greyish or brown pubescence, otherwise shoots, petioles

* Baehni (in Arch. Sci. Genève 17: 77 (1964) & 18: 31 (1965) & in Boissiera 11: 42 (1965)) disputes the generally accepted typification of *Lecomtedoxa* and adopts a new generic name, *Nogo*, for the species generally included there. This argument, however, fails to take into account article 41 of the present Intern. Code Bot. Nomencl. (1966), which allows indirect reference to the circumscription of a taxon described at a different rank. Thus although *Lecomtedoxa ogouensis* Dubard (1914) (now known as *Neolemonniera ogouensis* (Dubard) Heine) was the first species to be described in the genus, there is no need to reject *L. klaineana* (Pierre ex Engl.) Dubard (1915) (*Mimusops* (subgen. *Lecomtedoxa*) *klaineana* Pierre ex Engl. (1904)) as the type-species.

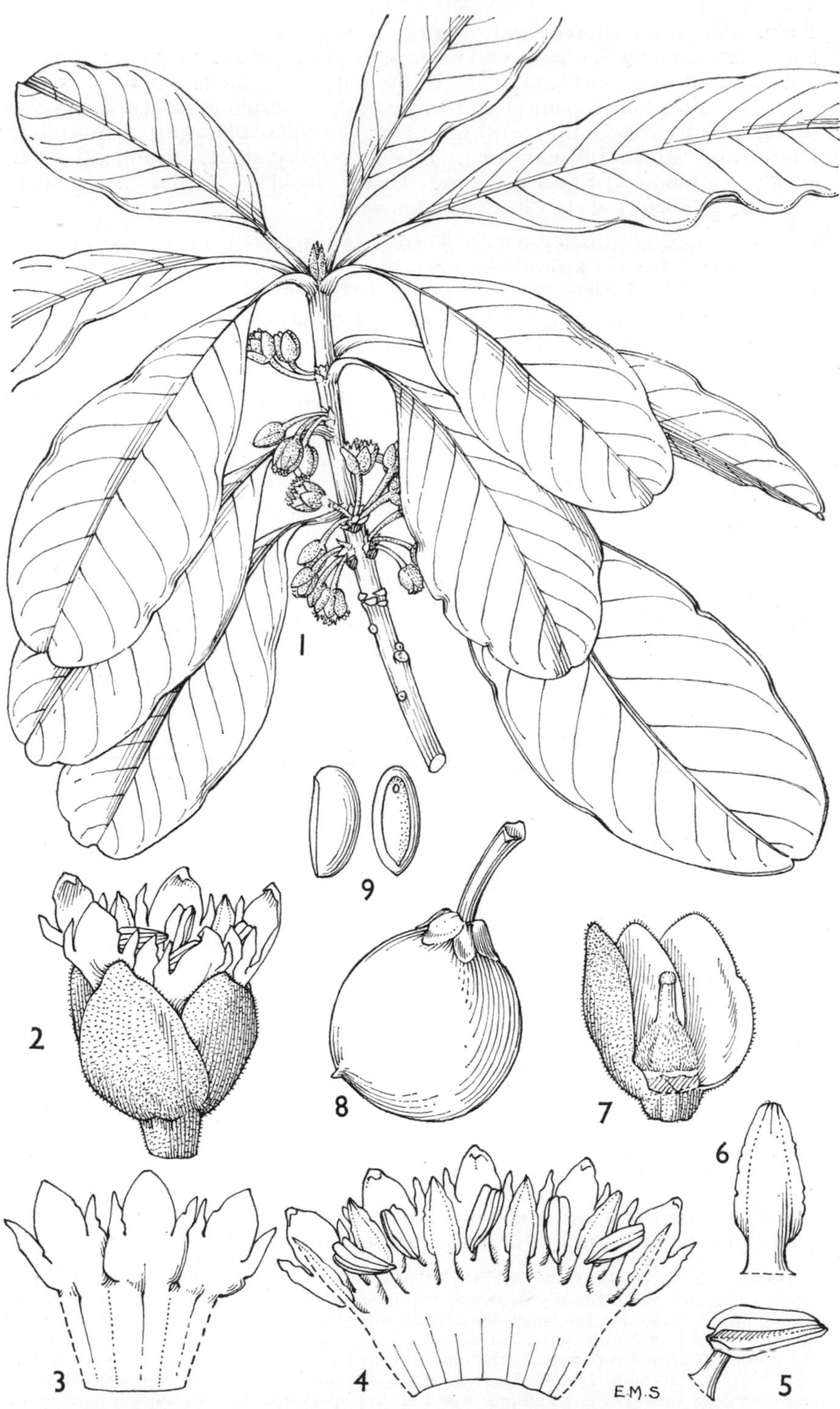

FIG. 9. *INHAMBANELLA HENRIQUESII*—**1,** flowering branch, × ⅔; **2,** flower, × 4; **3,** part of corolla, viewed from outside, × 4; **4,** corolla opened out, showing stamens and staminodes, viewed from inside, × 4; **5,** stamen, × 8; **6,** staminode, × 8; **7,** flower with two sepals and corolla removed to show gynoecium, × 4; **8,** fruit, × 1; **9,** seeds, × ⅔. 1–7, from *Rawlins* H25/58; 8, from *Gomes e Sousa* 4403; 9, after J. Sausotte-Guérel in Adansonia, mém. 1, t. 18/2.

and leaves glabrous. Petioles relatively long, up to 4 cm. Leaf-lamina obovate, elliptic-obovate to elliptic, usually 7–15(–18) cm. long, 3·5–8·5 cm. wide, stiffly coriaceous, apex obtuse to emarginate, cuneate, margin recurved, undulate and frequently irregular in shape; under-surface with primary lateral nerves 3–12 on each side of prominently raised midrib. Pedicels and calyx with greyish or brownish pubescence; pedicels up to 1·2 cm. long. Sepals broadly ovate-triangular, up to 5 mm. long, 4 mm. wide. Corolla-tube campanulate, up to 3 mm. long; lobes trifid; median segment ± ovate, up to 5 mm. long; lateral dorsal segments smaller and narrower. Stamens inserted at throat; filaments flattened, up to 3 mm. long; staminodes petaloid, broadly ovate, up to 3 mm. long. Ovary up to 2 mm. long; style up to 1 mm. long. Fruits red when mature, up to 3 cm. in diameter. Seeds up to 2·5 cm. long. Fig. 9.

KENYA. Kwale District: Mrima Hill, about halfway from Lungalunga to Msambweni, 16 Jan. 1964, *Verdcourt* 3936B!; Lamu District: Witu Forest, Pangani–Witu road, Feb. 1958 (fl.), *Rawlins* in *E.A.H.* 25/58! & NE. of Witu, 28 Feb. 1956, *Greenway & Rawlins* 8957!

TANGANYIKA. Ulanga District: Kilombero valley, Mikeregembe, 14 June 1958, *Ede & Amani* 3! & Mahenge, Dec. 1953, *R. Davies*!

DISTR. **K**7; **T**6; also coastal areas of Mozambique, extending up river valleys into Rhodesia and southern Malawi, and south to Natal

HAB. Lowland rain-forest and groundwater forest; 10–300 m.

SYN. *Mimusops henriquesii* Engl. & Warb., E.M. 8: 80, t. 25/A (1904), as "*henriquezii*", but corrected, l.c.: 88 (1904)
Lecomtedoxa henriquesii (Engl. & Warb.) Meeuse in Bothalia 7: 344 (1960) & in F.S.A. 26: 40 (1963)

VARIATION. The leaves of the East African material show a more regular arrangement of the primary lateral nerves and have a less distinctly reticulated nervation than those from Mozambique. This variation is accepted as part of a single species which may prove to be more widespread in the coastal areas than present records would indicate.

NOTE. The epithet of this species appears first without a description as *Mimusops henriquesiana* Engl. & Warb. in Bull. Soc. Étud. Colon. 10: 516 (1903) and as such has been followed by certain subsequent authors, but the spelling adopted at the place of first valid publication must be accepted.

10. BUTYROSPERMUM

Kotschy in Sitz. Akad. Wiss. Wien 50, Abth. 1: 357 (1865); Engl., E.M. 8: 22 (1904), *nom. conserv. propos.*

Vitellaria Gaertn. f., Fruct. 3: 131, t. 205 (1805); Baehni in Taxon 14: 42 (1965) & in Boissiera 11: 145 (1965)

Small to medium-sized tree. Leaves usually in terminal rosettes on thick much-scarred shoots, stipulate. Flowers axillary, normally clustered at apices of short-shoots. Sepals usually 8, arranged in two slightly dissimilar whorls, ± free to base. Corolla with short tube; lobes 8–10, irregularly denticulate. Stamens 8–10; filaments about same length as corolla, attenuate and bent near apex; anthers versatile, dehiscence extrorse. Staminodes present, petaloid, acute to cuspidate at apex, margins irregularly serrulate. Ovary subglobose, densely pilose, 8–10-locular; style long and slender; stigma simple. Fruit an ellipsoid or subglobose berry. Seeds large, solitary or sometimes 2–3, usually ± ovoid, often flattened on one side in fruits containing more than one seed; testa shiny brown with large pale lateral scar; endosperm absent; cotyledons plano-convex, thick and fleshy, containing abundant fat and oil.

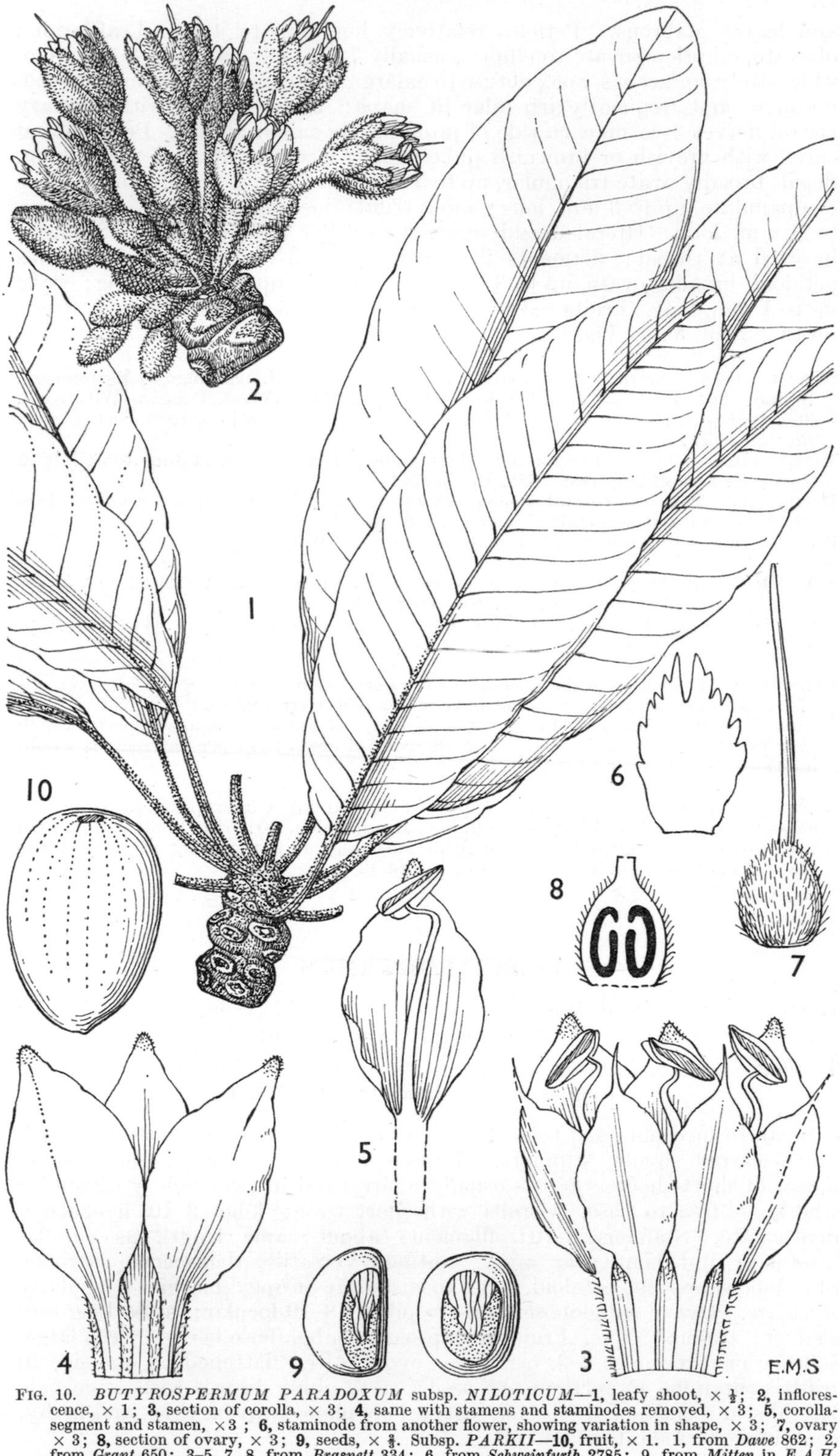

1 2 3 4 5 6 7 8 9 10 E.M.S

FIG. 10. *BUTYROSPERMUM PARADOXUM* subsp. *NILOTICUM*—**1,** leafy shoot, × ½; **2,** inflorescence, × 1; **3,** section of corolla, × 3; **4,** same with stamens and staminodes removed, × 3; **5,** corolla-segment and stamen, ×3 ; **6,** staminode from another flower, showing variation in shape, × 3; **7,** ovary × 3; **8,** section of ovary, × 3; **9,** seeds, × ⅔. Subsp. *PARKII*—**10,** fruit, × 1. 1, from *Dawe* 862; 2, from *Grant* 650; 3–5, 7, 8, from *Brasnett* 334; 6, from *Schweinfurth* 2785; 9, from *Mitten* in *E.A.H.* 97/51; 10, from *Coull* 1.

A monotypic genus restricted to the woodland and savannah regions of north tropical Africa.

B. paradoxum (*Gaertn. f.*) *Hepper* in Taxon 11: 227 (1962); Heine in F.W.T.A., ed. 2, 2: 21 (1963). Type: single seed of unknown origin (P, holo.)

Tree with stout bole and much-branched spreading crown, height up to 20 m.; bark usually grey or blackish, deeply fissured and splitting into squarish or rectangular corky scales. Short-shoots with conspicuous cycad-like annular leaf-base scars; young shoots, petioles and flower-buds with ferrugineous pubescence of varying density. Petioles long, one-third to half length of leaf-lamina. Leaf-lamina oblong to ovate-oblong, 10–25 cm. long, 4·5–14 cm. wide, rounded at apex, base acute to broadly cuneate, margin thickened and undulate; upper and lower surfaces of mature leaf glabrescent or puberulous; lateral nerves 20–30 each side, regularly and closely spaced, slightly arcuate. Flowers fragrant. Pedicels up to 3 cm. long, puberulous to densely pubescent. Outer sepals lanceolate, 9–14 mm. long, 3·5–6 mm. wide, pubescent to ± floccose externally; inner sepals slightly smaller. Corolla cream; tube 2·5–4 mm. long, glabrous or pilose externally; lobes broadly ovate, 7–11 mm. long, 4·5–7 mm. wide. Filaments 7–12 mm. long; anthers ± lanceolate, up to 4·5 mm. long. Staminodes up to 8 mm. long. Style 8–15 mm. long. Mature fruit greenish, up to 6·5 cm. long and 4·5 cm. in diameter, subglabrous or with pubescence persisting in small patches, containing a sweet pulp surrounding the seed. Seed up to 5 cm. long, 3·5 cm. in diameter. Fig. 10.

SYN. *Vitellaria paradoxa* Gaertn. f., Fruct. 3: 131, t. 205 (1805); Baehni in Boissiera 11: 146 (1965)

subsp. **niloticum** (*Kotschy*) *Hepper* in Taxon 11: 227 (1962). Type: Sudan Republic, Equatoria Province, Gondokoro, *Knoblecher* 61 (W, holo.)

Young shoots and leaves, pedicels and flower-buds with dense ferrugineous pubescence; old petioles and lower surface of older leaves puberulous or subglabrous but usually pubescent on midrib and nerves. Pedicels and outer sepals with dense floccose indumentum; outer sepals 10–14 mm. long, 5–6 mm. wide; inner sepals slightly smaller. Corolla-tube pilose externally; lobes 9·5–11 mm. long, 6·5–7 mm. wide. Filaments 10–12 mm. long. Style 12–15 mm. long. Fig. 10/1–9.

UGANDA. W. Nile District: Payida Escarpment, 21 Mar. 1945 (fr.), *Greenway & Eggeling* 7237! & 10 km. from Arua on Ayivu [Arivu] road, Dec. 1931 (fl.), *Brasnett* 334!; Teso District: Kumi, Feb. 1916 (fl.), *Snowden* 270!
DISTR. **U**1, 3, 4; eastern Congo and Sudan Republics
HAB. Grouped-tree and scattered-tree grassland, often protected and preserved in cultivated areas; frequently planted by African agriculturists but very rarely so outside the normal distribution range; 950–1500 m.

SYN. *B. niloticum* Kotschy in Sitz. Akad. Wiss. Wien 50, Abth. 1: 358, t. 1 (1865); F.P.S. 2: 373, fig. 138 (1952)
B. parkii (G. Don) Kotschy var. *niloticum* (Kotschy) Engl., E.M. 8: 23, fig. 9 (1904); Dubard in Ann. Mus. Col. Marseille, sér. 2, 10: 40 (1912); I.T.U., ed. 2: 388, fig. 78 (1952)
B. parkii (G. Don) Kotschy subsp. *niloticum* (Kotschy) J. H. Hemsl. in K.B. 15: 290 (1961)

NOTE. Subsp. *parkii* (G. Don) Hepper, widespread in the savannah regions of West Africa from Portuguese Guinea to the Central African Republic [Ubangi-Shari], has a less dense and shorter indumentum, also slightly smaller flowers, with the style, for instance, only 8–12 mm. long. There are probably a number of smaller varieties and morphological and physiological forms, the shape and size of both fruits and seeds and chemical analyses of the kernels appearing to vary from differing parts of the species range.

Subsp. *paradoxum* is a name restricted to the type, which cannot be certainly associated with either subsp. *parkii* or subsp. *niloticum*.

The tree is of considerable importance locally as a source of edible fat or vegetable butter extracted from the ripe seeds. A large volume of nuts, under the name shea nuts, is exported annually from West Africa; in Uganda, where the nuts are used on a smaller scale, a local market, developed during the Second World War, is being maintained, but no export trade has yet made an appearance.

11. MIMUSOPS

L., Sp. Pl.: 349 (1753) & Gen. Pl., ed. 5: 165 (1754); Lam in Bull. Jard. Bot. Buit., sér. 3, 7: 234 (1925)

Trees or shrubs. Stipules caducous. Leaf-lamina frequently elliptic to obovate; upper surface often glossy, lower surface practically glabrous or pubescent in young leaves; midrib raised, primary lateral nerves finely raised on both surfaces, secondary lateral nerves inconspicuous, venation reticulate. Flowers usually in axils of current leaves, pedicellate. Sepals 8, arranged in two dissimilar whorls of 4, ± free or slightly fused at base; inner sepals smaller and paler in colour than outer. Corolla of 8 members, fused into a short basal tube, upper parts further subdivided into three segments* giving 24 linear segments per flower. Stamens 8, epipetalous; anther dehiscence extrorse. Staminodes 8, alternating with stamens, usually simple, narrowly lanceolate or ligulate, densely pilose externally and along margins. Ovary usually 8-locular; ovules solitary, anatropous, with basal attachment. Fruit baccate, fleshy to ± coriaceous, 1–several-seeded, calyx persisting at base. Seeds ± ellipsoid, laterally compressed; testa horny and polished; scar ± basal to basi-ventral, usually (probably always in East Africa) small and scarcely longer than broad; abundant endosperm present; cotyledons flattened and foliaceous.

A genus found throughout the Old World tropics but absent from the American continent, with about 20 species in Africa.

M. elengi L., a SE. Asian species, which is a good shade-tree with sweetly scented small white flowers, has been cultivated at Amani and various coastal and near-coastal places in East Africa.

Flowering pedicels distinctly longer than petioles:
  Leaf-lamina usually obovate to elliptic-obovate, broadest in upper half; fruits subglobose:
    Leaves large, 7·5–13 cm. long; pedicels 2–4 cm. long; fruits up to 3·5 cm. in diameter . . . 1. *M. riparia*
    Leaves smaller, 3·5–9 cm. long; pedicels 1–2 (or rarely –2·5 cm.) long; fruits 1–2·5 cm. in diameter . . . . 2. *M. fruticosa*
  Leaf-lamina elliptic or oblong-elliptic to obovate-elliptic; flowering pedicels slender, 2–5 cm. long; fruits ovoid to ellipsoid with an acute or ± rostrate apex . . . . . . 3. *M. kummel*
Flowering pedicels shorter than petioles, or if a little longer then petioles not less than 1·5 cm. in length:
  Sepals ± elliptic to elliptic-oblong; flower-buds with bluntly rounded apices:
    Leaves and flowers clustered in terminal rosettes, often on dwarf shoots, younger leaves with soft dense ferrugineous indumentum . . . . 6. *M. schliebenii*

* See note page 2, concerning the nature and terminology of these segments.

Leaves not in terminal rosettes, younger leaves puberulous soon becoming glabrous . . *M. schimperi**
Sepals ovate to narrowly lanceolate; flower-buds ovoid and tapering to a subacute apex:
Pedicels 0·4–1 cm. long:
Leaves small, up to 5 cm. long, lanceolate to ovate . . . . 4. *M. acutifolia*
Leaves longer, 5–15 cm. long, elliptic to oblong-elliptic:
Indumentum of pedicel and calyx ochraceous to pale brown in colour (Uganda and Kenya uplands) . . . . . . 7. *M. bagshawei*
Indumentum of pedicel and calyx deep reddish-brown in colour (Kenya and Tanganyika lowlands) . . . . . . 8. *M. aedificatoria*
Pedicels 1–2·5 cm. long:
Leaf-lamina elliptic-oblong, 12–15 cm. long; pedicels curved through 90° with flowers pendulous . . . . 9. *M. penduliflora*
Leaf-lamina not as above, smaller, 4–11 cm. long; pedicels not markedly curving through 90° . . . . . . 5. *M. zeyheri*

1. **M. riparia** *Engl.* in E.J. 28: 448 (1900) & E.M. 8: 74, t. 28/C (1904); T.T.C.L.: 565 (1949); K.T.S.: 528 (1961). Type: Tanganyika, Kilosa/Iringa District, Uhehe on Ruaha R., *Goetze* 451 (B, holo. †, K, iso. !)

Small to medium-sized tree with spreading crown, height up to 20 m. Young branches and petioles with brownish puberulence, later becoming glabrous. Petioles 1–2·5 cm. long. Leaf-lamina subcoriaceous to coriaceous, elliptic to elliptic-obovate, (6–)7·5–13 cm. long, (3–)4–7·7 cm. wide, obtuse to shortly acuminate, broadly cuneate, glabrous; nerves and veins finely raised on both upper and lower surfaces. Flowers 1–3 in current leaf axils. Pedicels 2–4 cm. long; both pedicels and outer sepals with dense closely appressed indumentum of short brownish hairs. Outer sepals ± lanceolate, up to 1·5 cm. long and 5 mm. wide. Corolla cream; lobes trifid, narrowly lanceolate, up to 1·7 cm. long; tube up to 2·5 mm. long. Filaments up to 3·5 mm. long; anthers up to 4·5 mm. long; staminodes ± oblong to elliptic, up to 6 mm. long, apex attenuate or ± obtuse, densely pilose externally. Ovary subglobose, up to 3 mm. in diameter, pilose; style up to 2 cm. long. Fruit a pale yellow to reddish-yellow subglobose berry, up to 3·5 cm. in diameter; pericarp firm. Seeds 1–5, elliptic-oblong and laterally compressed, up to 1·9 cm. long; testa pale brown and glossy; scar ± basal.

KENYA. Teita District: Taveta, Lumi R., 22 Jan. 1936 (fr.), *Greenway* 4473 !
TANGANYIKA. Pangani District: between Hale and Makinjumbe, 1 July 1953 (fl. & fr.), *Drummond & Hemsley* 3132 ! & Mauri, 7 Sept. 1951 (fr.), *Greenway* 8701 !; Ulanga District: near Ifakara, May 1960 (fl. & yng. fr.), *Haerdi* 450/0 !
DISTR. **K7**; **T3**, 6–8; not recorded outside these districts
HAB. Riverine; 100–1000 m.

SYN. *M. useguhensis* Engl., E.M. 8: 67, t. 26/B (1904); T.T.C.L.: 566 (1949). Type: Tanganyika, Pangani District, Makinjumbe, *Scheffler* 258 (B, holo. †)
M. dependens Engl., E.M.: 69 (1904); T.T.C.L.: 565 (1949). Type: Tanganyika, Lindi District, Rondo [Mwera] Plateau, Liho R. at Palihope, *Busse* 2853 (B, holo. †, BM, EA, HBG, iso. !)

NOTE. A series of large-leaved and large-fruited riverine trees bearing close affinity with *M. fruticosa* A. DC. are brought together under the name *M. riparia*. Relation-

* See page 59.

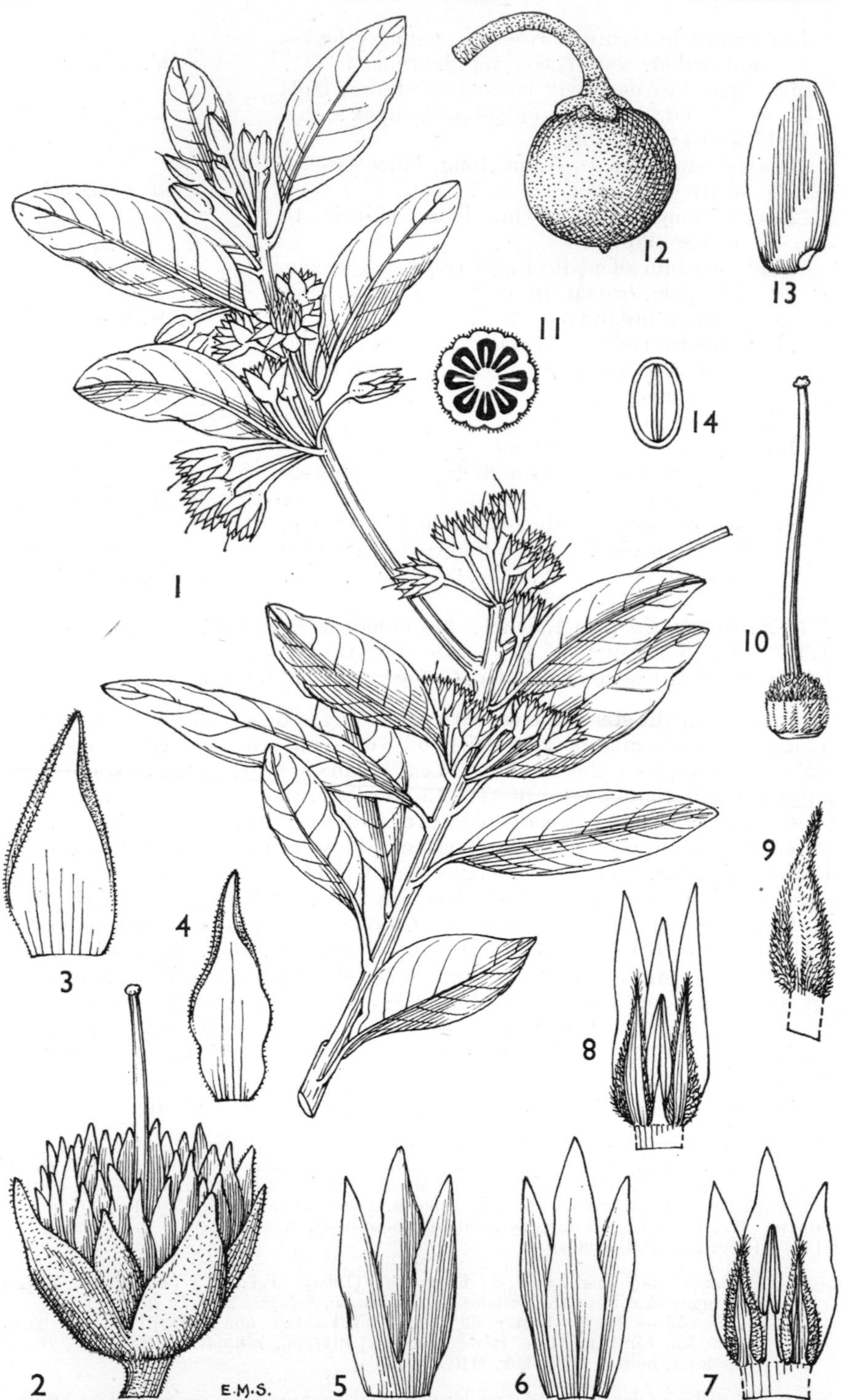

FIG. 11. *MIMUSOPS FRUTICOSA*—**1,** flowering branch, × ⅔; **2,** flower, × 4; **3,** outer sepal, × 4; **4,** inner sepal, × 4; **5,** section of corolla showing trifid lobe, outer aspect, × 4; **6,** same, inner aspect, × 4; **7, 8,** section of corolla with trifid lobe, stamen and staminodes, × 4; **9,** staminode, × 6; **10,** ovary, × 4; **11,** diagrammatic transverse section of ovary; **12,** fruit, × 1; **13,** seed, × 1½; **14,** transverse section of seed, × 1½. 1, from *Gillman* 774; 2–7, 9–11, from *Hughes* 81; 8, from *Vaughan* 1270; 12–14, from *Drummond & Hemsley* 3712.

ship between the two species is difficult to define, the obviously larger flowers and fruits of *M. riparia* confer upon the latter the appearance of a " giant " version, but few other characters except mere size exist upon which to base distinction. The species is maintained as such for the present, field investigation being required to further elucidate the true relationship between *M. fruticosa* and *M. riparia.*

2. **M. fruticosa** *A. DC.* in DC., Prodr. 8: 202 (1844); Baker in F.T.A. 3: 508 (1877); Engl., E.M. 8: 66, t. 23/B (1904); Dubard in Ann. Mus. Col. Marseille, sér. 3, 3: 50, fig. 18 (1915); T.T.C.L.: 565 (1949); U.O.P.Z.: 353 (1949); K.T.S.: 528 (1961). Type: *Bojer* specimen from plant cultivated in Mauritius, originally from East Africa (BM, prob. iso. !)

Shrub or much-branched small to medium-sized tree, height up to 20 m. Young branches and petioles with brownish puberulence, becoming glabrous. Petioles 0·5–1·5(–2·5) cm. long. Leaf-lamina coriaceous, elliptic-obovate to obovate, rarely ± broadly elliptic, 3·5–8 cm. long, 1·5–5·5 cm. wide, apex rounded to emarginate, rarely subacuminate, broadly to narrowly cuneate; upper surface dark glossy green, lower surface dull green, subglabrous or minutely puberulous with finely raised reticulate venation; lateral nerves ascending. Pedicels curved, 1–2·5 cm. long, with brownish pubescence. Outer sepals densely brownish pubescent, lanceolate, up to 9 mm. long. Corolla-lobes cream, trifid, linear-lanceolate, up to 8·5 mm. long; tube up to 2·5 mm. long. Filaments up to 2 mm. long; staminodes linear-lanceolate, up to 5 mm. long, densely pubescent externally. Ovary ± 2 mm. long, densely pilose; style slender and tapering, up to 1·2 cm. long. Mature fruit an orange or yellowish-orange to reddish globose to subglobose berry, 1–2·5 cm. in diameter; skin firm and tough. Seeds 1–5, rarely 6, obliquely elliptic or ± oblong, up to 2 cm. long and 5 mm. thick; testa glossy, deep brown, hard and horny; scar sub-basal. Fig. 11.

KENYA. Kilifi District: Mida, *C. W. Elliot* in *F.D.* 1499 !; Lamu District: Mkumbi [Mkumbe], 5 Apr. 1910 (fl.), *Battiscombe* 232 ! & S. bank of Tana R., 8 Nov. 1957 (fl. & fr.), *Greenway & Rawlins* 9481 !

TANGANYIKA. W. Usambara Mts. [NW. foot], Mkundi, *Gillman* 774 !; Tanga District: Umba R. E. of Mwakijembe, 12 Aug. 1953 (fr.), *Drummond & Hemsley* 3712 !; Rufiji District: Mafia, Chole I., 21 Sept. 1937 (fr.), *Greenway* 5286 !

ZANZIBAR. Zanzibar I., Mbweni, Feb. 1930 (fl.), *Vaughan* 1270 ! & NE. Mkunduchi, 27 Nov. 1930 (fr.), *Greenway* 2595 !; Pemba I., Jamvini, 12 June 1928, *Vaughan* 338 !

DISTR. **K**7; **T**3, 6, 8; **Z**; **P**; Mozambique and Rhodesia, also Comoro Is. and Madagascar

HAB. Lowland dry evergreen forest, riverine forest and coastal evergreen thickets; 0–750 m.

SYN. *M. fruticosa* Bojer, Hort. Maurit.: 198 (1837), *nomen subnudum*
M. kirkii Baker in F.T.A. 3: 507 (1877); Engl., E.M. 8: 67 (1904). Types: Mozambique, Lower Shire valley, Shamo, *Kirk* (K, syn. !, four gatherings, various dates)
M. kilimanensis Engl., E.M. 8: 67 (1904); T.T.C.L.: 566 (1949). Type: Mozambique, Zambezia, near Quelimane [Kilimane], Puguruni, *Stuhlmann* 1007 (B, holo. †, HBG, iso. !)
M. usambarensis Engl., P.O.A. C: 307 (1895), pro parte, emend. Engl., E.M. 8: 74, t. 29/B (1904); T.T.C.L.: 565 (1949). Type: Tanganyika, Tanga District, Moa, *Holst* 3043 (HBG, lecto. !, K, isolecto. !)
M. usaramensis Engl., E.M. 8: 66 (1904); T.T.C.L.: 566 (1949). Type: Tanganyika, Dar es Salaam, *Schlechter* (B, holo. †)
M. busseana Engl., E.M. 8: 79 (1904); T.T.C.L.: 566 (1949). Type: Tanganyika/Mozambique, on middle reaches of R. Ruvuma near Kwa Mtora, *Busse* 1025 (B, holo. †, EA, iso. !)
[*M. zeyheri* sensu Meeuse in Bothalia 7: 361 (1960) & in F.S.A. 26: 47 (1963), pro parte, quoad syn. *M. kirkii*, *non* Sond.]

VARIATION. A very variable species closely related to *M. zeyheri* Sond. The species in the present sense has been taken to include a series of taxa forming a well-defined coastal distribution pattern from Kenya to Mozambique and SE. Rhodesia, with outliers in the Comoro Is. and Madagascar. Petioles in the southern part of the range tend to be long in relation to the lamina and extreme forms contrast with the short petioles of material from Kenya and Tanganyika. Occasional long-petioled specimens, however, have been collected in Tanganyika. The texture and shape of the leaves are variable; stiffly coriaceous leaves are usually to be found on shrubs from the coastal thickets, and those of a more chartaceous nature from trees in riverine forest. See also under 5, *M. zeyheri*.

3. **M. kummel** *A. DC.* in DC., Prodr. 8: 203 (1844); Baker in F.T.A. 3: 508 (1877); Engl., E.M. 8: 75, t. 30/A (1904); I.T.U., ed. 2: 400, fig. 83 (1952); Aubrév., Fl. For. Côte d'Ivoire, ed. 2, 3: 124, t. 295 (1959); K.T.S.: 528 (1961); Heine in F.W.T.A., ed. 2, 2: 20 (1963). Type: Ethiopia, Tigre, Mt. Scholoda, *Schimper* 280 (K, lecto.!, BM, OXF, isolecto.!)

Small to medium-sized tree, or shrub, height up to 25 m. Young branches and petioles with ferrugineous or brownish pubescence, becoming glabrous. Petioles 5–15 (rarely –30) mm. long. Leaf-lamina coriaceous, elliptic to oblong-elliptic or obovate-elliptic, 4·5–12 cm. long, 2·5–5 cm. wide, apex shortly acuminate, sometimes obtuse or emarginate, narrowly to broadly cuneate; upper surface dark glossy green, lower surface paler with slightly raised reticulate venation, practically glabrous or with scattered hairs along midrib. Flowers fragrant, usually 2–4 per axil; pedicels slender, curved, (1·4–)2–5 cm. long, densely ferrugineous pubescent. Outer sepals ± lanceolate, 9·5–12 mm. long, densely ferrugineous pubescent externally. Corolla creamy-white; lobes basically trifid but two outer segments sometimes split into two (rarely three) linear segments; segments 9–12 mm. long; tube up to 2 mm. long. Filaments 2–3·5 mm. long; staminodes linear-lanceolate, 4–6 mm. long, apex attenuate, densely pubescent externally. Ovary densely covered with long straight hairs; style slender, tapering, 1–1·2 cm. long. Fruit an orange or orange-red ellipsoid to ovoid berry, up to 2·5 cm. long, apex obtuse, acute or shortly rostrate. Seed solitary, ellipsoid, up to 1·8 cm. long; testa brown, hard and horny; scar obliquely basal.

UGANDA. W. Nile District: Payida [Paida], Mar. 1935 (fl.), *Eggeling* 1923 in *F.D.* 1683!; Teso District: Serere, Mar. 1932 (fl.) & Feb. 1933 (fl. & fr.), *Chandler* 617! & 1083!; Mbale District: [NE. Elgon], Bukwa, 14 Apr. 1927 (fl.), *Snowden* 1066!

KENYA. Baringo District: Kamasia, Tarambas Forest, Nov. 1930 (fr.), *Dale* in *F.D.* 2446!; Kiambu District: Ngong Forest, Feb. 1921 (fl. & fr.), *Gardner* in *F.D.* 1147!; Machakos District: Momandu area, Sept. 1932 (fr.), *Gibbons* in *F.D.* 2899!

TANGANYIKA. Mwanza District: Geita, 9 July 1953 (fl.), *Tanner* 1575!; Musoma District: Wogakuria Hill, 30 Dec. 1964 (fl.), *Greenway & Turner* 12002!; Rungwe massif, upper slopes of Kibila, 27 Oct. 1913 (fr.), *Stolz* 2273!

DISTR. **U**1–3; **K**1, 3–7; **T**1–4, 6–8; Eritrea, Ethiopia, Sudan Republic, S. Sahara to West Africa

HAB. Widespread in riverine forest and riparian vegetation, often found as scattered small trees, also in upland dry evergreen forest (Nairobi area and Kenya highlands), in wooded grasslands and on rocky hills; 500–2100 m.

SYN. *Imbricaria fragrans* Baker in F.T.A. 3: 509 (1877). Type: S. Nigeria, Yoruba, *Barter* 1217 (K, holo.!)

Mimusops longipes Baker in K.B. 1895: 149 (1895); Engl., E.M. 8: 74, t. 28/A (1904). Type: Nigeria, W. Lagos, *Rowland* (K, holo.!)

M. kilimandscharica Engl., E.M. 8: 68, t. 27/A (1904); T.T.C.L.: 566 (1949). Types: Kenya, without precise locality, *C. F. Elliot* 100 (B, syn. †) & Tanganyika, Moshi District, Kahe, *Volkens* 2192 (B, syn. †, BM, isosyn.!)

M. langenburgiana Engl., E.M. 8: 70, t. 28/D (1904); T.T.C.L.: 566 (1949). Type: Tanganyika, Rungwe District, near Tukuyu [Langenburg], *Goetze* 864 (B, holo. †)

M. djurensis Engl., E.M. 8: 75, t. 30/B (1904); F.P.S. 2: 375 (1952). Types: Sudan Republic, Bahr el Ghazal, *Schweinfurth* 1379, 1527 & 2428 (all K, isosyn.!)

M. pohlii Engl., E.M. 8: 76, t. 30/C (1904); T.T.C.L.: 565 (1949). Type: Tanganyika, Mwanza District, Kayenzi [Kagehi], *Fischer* 477 (B, holo. †)

M. fragrans (Baker) Engl., E.M. 8: 77, t. 31/A (1904); Dubard in Ann. Mus. Col. Marseille, sér. 3, 3: 51, fig. 20 (1915); F.W.T.A. 2: 14 (1931); Aubrév.,

Fl. For. Côte d'Ivoire 3: 104, t. 281 (1936); I.T.U.: 227, fig. 67 (1940); F.P.S. 2: 375 (1952); Fl. Cameroun 2: 35, t. 4/1–6 (1964)
M. kerstingii Engl., E.M. 8: 78, t. 26/D (1904). Type: Togo Republic, Sokodé, *Kersting* (K, iso. !)
M. stenosepala Chiov. in Atti R. Accad. Ital. 11: 47 (1940). Type: Ethiopia, Borana, Neghelli, *Senni* 1015 (FI, holo. !)
M. sp. near *M. warneckei* Engl. sensu I.T.U., ed. 2: 402 (1952)

Note. The species *M. kummel* as here defined brings together a wide range of geographical variants which have proved impossible to key out as distinct taxa. A long transitional series exists between the large leaved and flowered end of the range in West Africa, with flower-buds densely ferrugineous floccose externally and the outer whorl of corolla-segments tending to be further split up each into three or more laciniate parts, and the smaller leaves and flowers with pubescence less floccose and the outer segments tending to be entire as in Kenya and in some Tanganyika material. Within this range it is possible to find much variation of the characters mentioned. Outer corolla-segments are split for only half their length as in *Bally* 6241 from near Nairobi, or taking the two gatherings *Chandler* 617 and *Snowden* 1066 both from Uganda, agreeing closely in details of flower structure and dimensions and also in vegetative characters, but the outer corolla-segments of the former are simple whereas those of the latter are regularly trifid. Extending southwards through Mozambique to South Africa is another related taxon *M. obovata* Sond., which has been maintained at specific level for the present but which appears to be closely allied to the *M. kummel* complex.

4. **M. acutifolia** *Mildbr.* in N.B.G.B. 14: 108 (1938); T.T.C.L.: 565 (1949). Type: Tanganyika, Lindi District, Lake Lutamba, *Schlieben* 6102 (B, holo. †, BM, K, iso. !)

Shrub or small tree up to 15 m. high with slender branchlets, pubescent when young, later becoming glabrous. Petiole 5–10 mm. long. Leaf-lamina rigidly coriaceous, lanceolate to ovate, up to 5 cm. long and 2·2 cm. wide, gradually tapering into a long acuminate apex, base obtuse to broadly cuneate, nervation inconspicuous, primary nerves ± 10 and slightly raised on lower surface. Flowers in axils of current leaves; pedicels slender, curving, up to 6 mm. long, ferrugineous pilose. Outer sepals up to 6 mm. long and 2 mm. wide. Corolla white; lobes trifid; median segment up to 4 mm. long; outer segments up to 1·5 mm. long; tube short, up to 1·5 mm. long. Stamens 3; filaments ± 1 mm. long; anthers up to 2 mm. long; staminodes subequal in length to corolla-lobes. Ovary subglobose, tapering into a 6 mm. long style. Fruit and seeds unknown.

Tanganyika. Lindi District: Noto Plateau, Lake Lutamba, 9 Mar. 1935 (fl.), *Schlieben* 6102!
Distr. **T** ?3, 8; not known elsewhere
Hab. "Bushland"; ± 450 m.

Note. The species is known with certainty from a single collection. To this have been added two further gatherings from the E. Usambara region, *Greenway* 5852! from Mlinga Peak and *Zimmerman* in *Herb. Amani* G7671! from Monga, which, although by no means good matches, show enough characters in common to suggest ecotypic variation and to imply further that the type may indeed represent a reduced form from an extreme habitat. This addition would bring the taxon close to a series of specimens from South Africa and Mozambique, known as *M. obovata* Sond. (*M. woodii* Engl., *M. rudatisii* Engl. & Krause). *M. acutifolia* Mildbr. may well prove to be a northern extension of this species. Both *M. obovata* and *M. acutifolia* are to be compared in turn with the Kenya form of *M. kummel*, and all may be regarded as part of a longer series of closely related taxa varying from the form previously known as *M. fragrans* in West Africa, with gradual transitions in leaf-size and form, pedicel-length and flower-size, and to a lesser extent fruit-shape and size, to the other end of the scale in South Africa. For the present *M. acutifolia* Mildbr. is maintained at specific level, but with little conviction, and differs from other East African *Mimusops* species by the slender twiggy branchings and comparatively small leaves with the lamina tapering into a long acuminate apex.

5. **M. zeyheri** *Sond.* in Linnaea 23: 74 (1850); Engl., E.M. 8: 73, t. 27/C (1904); Meeuse in Fl. Pl. Afr. 30, t. 1164 (1954) & in Bothalia 7: 361 (1960), pro parte, excl. syn. *M. kirkii*; F.F.N.R.: 321 (1962); Meeuse in F.S.A. 26: 47 (1963), pro parte, excl. syn. *M. kirkii*. Type: South Africa, Transvaal, Magaliesberg Range, *Zeyher* 1130 (LY, holo. !, K, OXF, iso. !)

Small to medium tree (rarely a shrub), height up to 20 m., with much branched dense evergreen crown; bark grey or blackish, rough. Young branches and petioles with ferrugineous pubescence, later wearing away. Petioles 1–3 cm. long. Leaf-lamina thinly coriaceous to coriaceous, ± elliptic, obovate-elliptic to ± oblanceolate, (4–)5–9(–11) cm. long, (1·7–)2·3–4(–4·5) cm. wide, obtuse to ± broadly acute or rarely shortly acuminate, broadly cuneate; upper surface dark green and glossy, with slightly raised nervation, lower surface paler, practically glabrous except on and near midrib; lateral nerves ascending, finely raised. Flowers usually 2–3 per axil; pedicels 1·5–2·5(–3·5) cm. long, curved, with dense ferrugineous or brownish pubescence. Outer sepals ± lanceolate, up to 11 mm. long, 4 mm. wide, externally with dense ferrugineous or brownish pubescence. Corolla-lobes cream or whitish, trifid; segments linear-lanceolate, up to 9·5 mm. long; tube up to 2 mm. long. Filaments up to 3 mm. long. Staminodes narrowly lanceolate, up to 5 mm. long, narrowly acute, simple or sometimes minutely laciniate. Ovary up to 2·5 mm. long, densely pilose; style tapering, up to 1 cm. long. Mature fruit an orange or yellowish ± ellipsoid or ovoid-subglobose berry, 1·5–3·5 cm. long and 1·5–2·5 cm. in diameter, with firm tough skin. Seeds 1–2 (very rarely more), variable in shape, generally ± ellipsoid or sometimes ± oblong when more than one seed present, 1–2(–2·5) cm. long, up to 1·2 cm. wide and 6 mm. thick; testa shiny pale brown, hard and horny; hilum ± basal in small circular depression.

TANGANYIKA. Ufipa District: Sumbawanga–Pito, 25 Nov. 1949 (fl.), *Bullock* 1935!; Dodoma District: Saranda–Kondoa road, Ruwiri R., 30 Nov. 1932 (fl.), *B. D. Burtt* 3807!; Mbeya District: Mbeya–Sao Hill road, Kinani R., 29 Oct. 1947 (fl.), *Brenan & Greenway* 8232!

DISTR. T4–7; Mozambique, Malawi, Zambia, Rhodesia and Botswana to Natal

HAB. Riverine vegetation and thicketed ravines in *Brachystegia*-woodland country; 1150–2100 m.

SYN. *M. monroi* S. Moore in J.B. 49: 154 (1911). Type: Rhodesia, Fort Victoria, *Monro* 761 (BM, holo. !)

NOTE. Closely related to and, over part of the range, difficult to distinguish from *M. fruticosa* A. DC. In East Africa, however, the two species are reasonably well defined both ecologically and geographically. Coupled with this are the ± elliptic to oblanceolate leaf-shapes and the ovoid to ellipsoid fruits of *M. zeyheri*. *M. fruticosa* has been treated as a distinct coastal species in the present work, see page 53.

6. **M. schliebenii** *Mildbr. & G. M. Schulze* in N.B.G.B. 12: 197 (1934); T.T.C.L.: 565 (1949); K.T.S.: 528 (1961). Type: Tanganyika, near Kilwa, *Schlieben* 2520 (B, holo. †, BM, iso. !)

Medium-sized tree, height up to 20 m. with repeated subapical branching; young shoots ferrugineous pubescent later becoming glabrous. Leaves clustered at shoot apices; petioles 1–2 cm. long, with ferrugineous pubescence. Leaf-lamina coriaceous, elliptic to obovate-elliptic, (3·5–)4·5–11 cm. long, (2–)2·5–4·5 cm. wide, acute to obtuse, broadly cuneate; upper surface glabrous with finely raised nervation, lower surface with soft dense ferrugineous indumentum when young, later stripping away and becoming glabrous. Flowers clustered in axils of current or fallen leaves. Pedicels 7–15 mm. long, ferrugineously pubescent. Outer sepals elliptic to elliptic-oblong, 6–8 mm. long, up to 3 mm. wide, with dense short ferrugineous pubescence. Corolla-lobes trifid, but with outer pair of segments further split into two (sometimes incompletely); median segment narrowly ligulate, up to 6 mm. long; outer segments broader and about same length; tube up to 1·5 mm. long. Filaments up to 1·5 mm. long; anthers up to 3 mm. long. Staminodes narrowly lanceolate, up to 3 mm. long. Ovary up to 2 mm. long, tapering into a short style up to 2·5 mm. long. Fruit ovoid, up to 2·9 cm. long and 1·7 cm. in diameter.

KENYA. Kwale District: Samburu, 12 Feb. 1953 (fl.), *Bally* 8557!
TANGANYIKA. Tanga District: between Moa and Mwakijembe, Feb. 1950 (fl.), *Kemode* in *Herb. Amani* 9931!; Kilwa District: near Kilwa, Mkindu, 25 June 1932 (fl. & fr.), *Schlieben* 2520!
DISTR. **K7**; **T3**, 8; not known outside this area
HAB. Coastal woodlands and evergreen bushlands; 150–300 m.

NOTE. This distinctive but poorly represented species has a growth habit easily distinguishing it from other East African *Mimusops* species. Flowers and fruits are borne near the apices of either short or long shoots in a manner reminiscent of *Manilkara mochisia* (Baker) Dubard, a species of similar habitat, and the growth is frequently continued by a succession of repeated subapical branchings producing a zigzag pattern.
The description of fruits has been taken from the original species description, none being available in herbaria.

7. **M. bagshawei** *S. Moore* in J.B. 44: 86 (1906); I.T.U., ed. 2: 400 (1952); K.T.S.: 527 (1961). Type: Uganda, Mengo District, Entebbe, *Bagshawe* 684 (BM, holo.!)

Tall tree with long straight bole, height up to 40 m.; bark of twigs deep purplish-brown, rough and fissured. Young branches and petioles puberulous, becoming glabrescent. Petioles 1–2(–2·3) cm. long. Leaf-lamina elliptic to elliptic-oblong or rarely obovate-oblong, 5·5–12(–14) cm. long, 2–5·3(–5·6) cm. wide, apex shortly acuminate or ± cuspidate, cuneate; upper surface with slightly raised nerves, lower surface practically glabrous or with scattered hairs along midrib; lateral nerves and veins slightly raised. Flowers fragrant, 2–4 per axil, ± pendulous; pedicels curved, up to 1 cm. in length, with dense buff-coloured pubescence. Outer sepals ± ovate, 5·5–7 mm. long, with dense buff to pale brown pubescence externally. Corolla greenish-yellow or cream; lobes trifid; segments oblong-lanceolate to narrowly lanceolate, up to 5 mm. long, with outer pair sometimes irregularly serrate near apex (very rarely bifid to base); tube up to 1·5 mm. long. Filaments up to 1·5 mm. long; staminodes oblong-lanceolate, up to 4·5 mm. long, acute. Ovary densely pilose with short pale brown hairs; style tapering, up to 4 mm. in length. Fruit a yellow or orange-yellow broadly ovoid berry, up to 2·5 cm. long and 2 cm. in diameter, with greyish flannel-like indumentum when young, later becoming glabrous. Seeds 1–3, up to 9 mm. long, tending to be triangular in cross-section, obscurely ridged along the broader margin, and usually with a small basal projection; testa brown; scar ± basal.

UGANDA. Bunyoro District: Budongo Forest, Dec. 1933 (yng. fl.), *Eggeling* 1474 in *F.D.* 1419!; Ankole District: Kalinzu Forest, July 1938 (fr.), *Eggeling* 3759!; Mengo District: 36 km. on [Kampala–]Entebbe road, Oct. 1937 (fl.), *Chandler* 1986!
KENYA. Ravine District: Maji Mazuri, *T. A. Angus* in *F.D.* 3215!; Kericho District: Sotik, Kibajet Estate, 6 Oct. 1948 (fr.), *Bally* 6470!; Masai District: Lolgorien, Oct. 1935 (fl.), *Oates* in *F.D.* 3363!
TANGANYIKA. Bukoba District: Minziro Forest, 16 Apr. 1956 (fl. bud), *Hughes* 351! & Feb. 1959 (fl.), *Procter* 1166!
DISTR. **U2–4**; **K3**, 5, 6; **T1**; occurs also in southern Sudan Republic
HAB. Lowland and upland rain-forest, found in both well-distributed and irregularly distributed rainfall types of the latter; 1100–2400 m.

SYN. *M. ugandensis* Stapf in J.L.S. 37: 523 (1906); I.T.U.: 227 (1940); F.P.S. 2: 375 (1952). Type: Uganda, Masaka District, Buddu, *Dawe* 252 (K, holo.!)
M. ugandensis Stapf var. *heteroloba* Stapf in J.L.S. 37: 524 (1906). Types: Uganda, Bunyoro District, Bugoma Forest, *Dawe* 724 & Toro District, Kibale Forest, *Dawe* 509 (both K, syn.!)

NOTE. Another species *M. andongensis* Hiern (*M. warneckei* Engl.) from West Africa and Angola appears to be a close relative. The main point of difference lies in the colour of bark of young twigs, being pale grey in *M. andongensis* and deep purplish-brown in *M. bagshawei*. Correlated with this is a small difference in staminode shape; the narrowly filiform apex of *M. andongensis* is much longer than the blunt to narrowly acute apex of *M. bagshawei*. Good flowering material, however, is scanty and it may be imprudent to place too great a significance upon the latter character. Gatherings of *M. bagshawei*, including *Bally* 2693!, 18 Aug. 1943, and 4630!, Aug. 1945, have been obtained from the Nairobi Arboretum where the species is in all probability cultivated. A note on *Bally* 4630 suggests the tree to have been introduced from Kakamega.

8. **M. aedificatoria** *Mildbr.* in N.B.G.B. 14: 109 (1938); T.T.C.L.: 565 (1949); K.T.S.: 527 (1961). Type: Tanganyika, NW. side of Uluguru Mts., *Schlieben* 3896 (B, holo. †, BM, K, iso.!)

Medium-sized tree, height up to 30 m., slightly buttressed at base. Young branches and petioles with dense ferrugineous or dark brown pubescence, older branches glabrescent. Petioles 1·5–3 cm. long. Leaf-lamina oblong-elliptic to oblanceolate, 6–15·5 cm. long, 1·8–5·1 cm. wide, shortly acuminate, rarely acute, base tapering and narrowly cuneate; young leaves with dense ferrugineous pubescence on both surfaces, soon rubbing away; upper surface with fine raised and reticulate venation, lower surface with less prominent venation, practically glabrous or with hairs along midrib. Flowers 2–4 per axil; pedicels slightly curved, 5–10 mm. long, with dense ferrugineous hairs. Outer sepals ± ovate, up to 6 mm. long, densely ferrugineous pubescent externally. Corolla cream; lobes trifid; median lobe wider than two outer, up to 5·5 mm. long; tube up to 2 mm. long. Filaments up to 1·5 mm. long; staminodes ± lanceolate, up to 4 mm. long, acute, densely hairy externally. Ovary with dense straight hairs; style 1·5–3 mm. in length. Mature fruit a yellowish subglobose to broadly ellipsoid berry, up to 2 cm. long and 1·5 cm. in diameter; pericarp thick and leathery. Seed solitary, pale brown, up to 1·5 cm. long, ± triangular in cross-section and ridged along the broader margin with a small projection at base; scar basal.

KENYA. Kwale District: Shimba Hills, Cha Shimba Forest, 22 July 1953 (fr.), *Templer* 3! & near Kwale in Cha Shimba Forest, Apr. 1938 (fl.), *Dale* in *F.D.* 3869!
TANGANYIKA. Lushoto District: Amani, Kiumba, 25 Mar. 1913 (fl.), *Lommel* in *Herb. Amani* 3968! & Mashewa, 19 Dec. 1938 (fl.), *Chambo* 10!; Morogoro District: Mtibwa Forest Reserve [near Turiani], 20 Aug. 1951 (fr.), *Greenway & Farquhar* 8624!
DISTR. **K7**; **T3**, 6, 7; and southwards into Malawi
HAB. Lowland rain-forest, riverine forest and groundwater forest; 200–1600 m.

NOTE. Relationships between this species and *M. blantyreana* Engl. are not clearly understood. Type material of the latter has not been seen but there is reason to believe the two to be closely related. Fruits of *M. aedificatoria* are small, ± ellipsoid or subglobose, and are borne on short stalks, ± 1 cm. long. In leaf characters there is close similarity between the two, the leaf shapes of *M. aedificatoria* falling near to some of those Malawi specimens thought to agree with the author's original description of *M. blantyreana*. Present material, particularly of flowering and fruiting stages, is inadequate, however, and further action is not thought justifiable at the present time.

9. **M. penduliflora** *Engl.* in E.J. 28: 448 (1900) & E.M. 8: 69, t. 28/B (1904); T.T.C.L.: 566 (1949). Type: Tanganyika, S. Uluguru Mts., *Goetze* 342 (B, holo. †, K, iso.!)

Tree up to 10 m. in height; young branches, petioles, pedicels and flower-buds with dense ferrugineous indumentum. Petioles up to 3 cm. long. Leaf-lamina elliptic-oblong, 12–15 cm. long, 4–5 cm. wide, ± obtuse, broadly cuneate, subcoriaceous, glabrous and shiny on upper surface; primary lateral nerves ± 15 each side, finely raised on both surfaces. Flowers clustered in axils of current leaves, pendulous. Sepals narrowly lanceolate, acuminate, up to 10 mm. long and 4 mm. wide. Corolla-segments linear-lanceolate, up to 7 mm. long. Filaments up to 2·5 mm. long; anthers up to 4 mm. long; staminodes narrowly lanceolate and ± equal to length of stamens, externally pilose. Ovary densely pilose, tapering into a slender style up to 1 cm. long.

Tanganyika. Morogoro District: S. Uluguru Mts., rocky bank of the Mbakana [Mbakano] R., Dec. 1898 (fl.), *Goetze* 342!
Distr. T6; not known elsewhere
Hab. Probably riverine forest; 600 m.

Note. Known only from a single type-specimen which may perhaps represent an extreme form of one of the more widespread species. Further gatherings from the type area possibly showing a range of flowering and fruiting material are needed in order to clarify the taxonomic standing of the species.

Doubtful species

M. schimperi *A. Rich.*, Tent. Fl. Abyss. 2: 22 (1851); Baker in F.T.A. 3: 507 (1877); Engl., E.M. 8: 76 (1904), pro parte. Types: Ethiopia, Tigre, Tacazze, *Schimper* 697 & 873 (both K, OXF, isosyn.!)

A small tree with spreading crown, height up to 15 m., distinguished from related *Mimusops* species by the relatively long petioles, 3·5–5 cm. in length and usually about half the length of the lamina. The flower-buds are short and blunt with the outer sepals presenting a striped appearance due to their paler discolorous margins.

Note. Two gatherings have been seen, *Snowden* 1895, Jan. 1930, and *Maitland* 696, Apr. 1923, both from the Entebbe Botanic Gardens. It is probable that the species has been planted here. It occurs naturally in the Sudan Republic, Ethiopia and Arabia and may well be present within the area covered by the present work. See also I.T.U., ed. 2: 400 (1952).

12. VITELLARIOPSIS

(Baill.) Dubard in Ann. Mus. Col. Marseille, sér. 3, 3: 44 (1915); Aubrév. in Adansonia 3: 41 (1963)

[*Butyrospermum* sensu Baker in F.T.A. 3: 504 (1877), *non* Kotschy]
Mimusops L. sect. *Vitellariopsis* Baill. in Bull. Soc. Linn. Paris 2: 942 (1891) & Hist. Pl. 11: 270 (1891); Engl., E.M. 8: 80 (1904)
Austromimusops A. Meeuse in Bothalia 7: 347 (1960)
[*Baillonella* sensu Baehni in Boissiera 11: 120 (1965), pro parte, *non* Pierre]

Shrubs or small trees. Leaves terminal and crowded at shoot apices. Stipules present, sometimes persistent and bristle-like. Leaf-lamina elliptic to obovate or oblanceolate, coriaceous, glabrous or ± puberulous especially when young; midrib and lateral nerves raised, primary lateral nerves ascending, secondary nerves inconspicuous, venation reticulate. Flowers clustered in current leaf axils, usually several per axil, long-pedicellate. Sepals 8, arranged in two dissimilar whorls of 4, slightly fused at base, densely pubescent externally. Corolla of 8 members fused at base into a short tube, each member usually divided again into three segments. Stamens 8, alternating with 8 broad entire staminodes. Ovary densely hairy, 8-locular; ovules solitary. Fruit baccate, dryish and coriaceous, 1–few-seeded. Seeds large, ± ovoid to subglobose, but frequently flattened on one side in fruits having more than one seed; scar large and covering up to half seed surface;

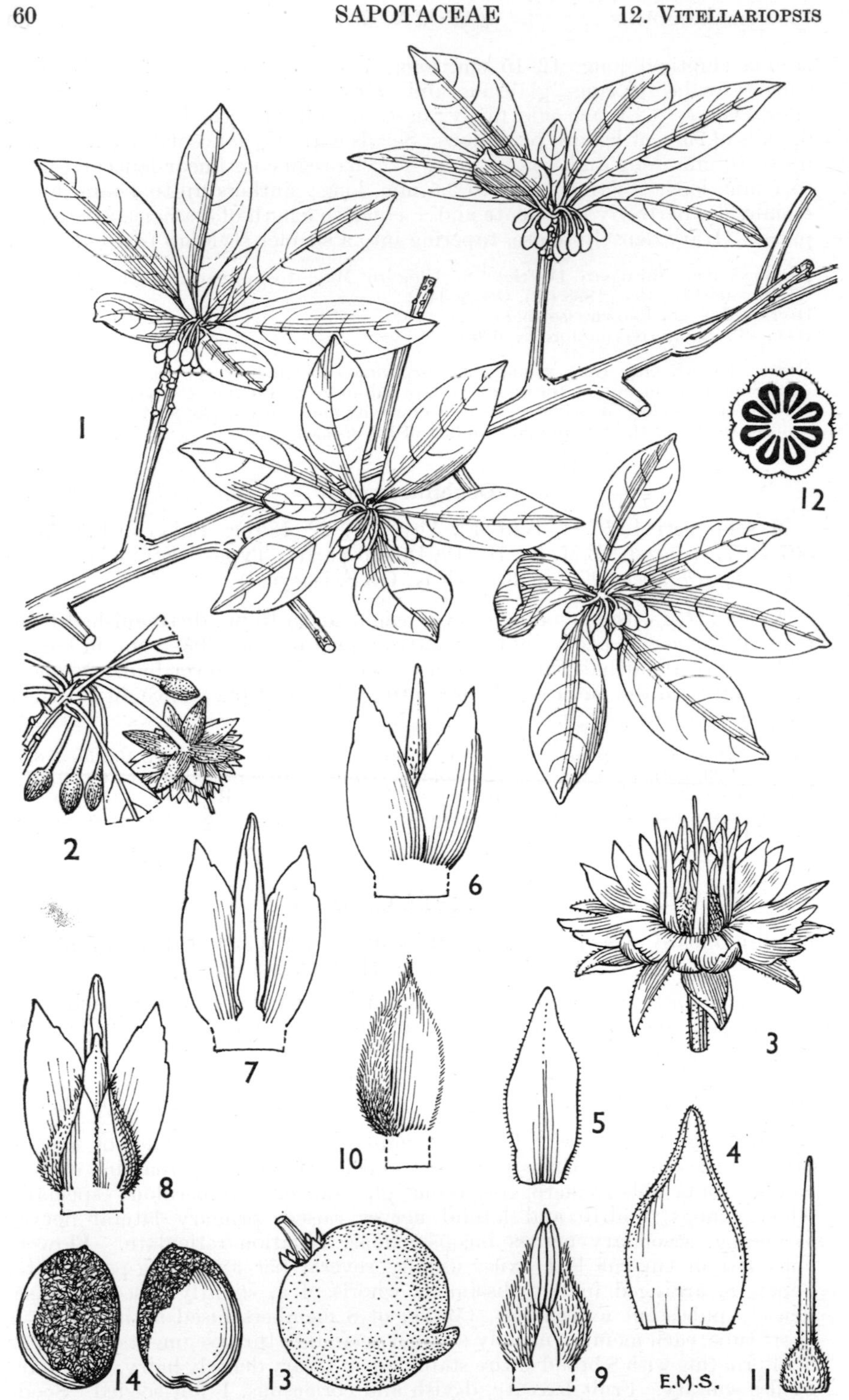

FIG. 12. *VITELLARIOPSIS KIRKII*—**1,** flowering branch, × ⅔; **2,** part of inflorescence, × 1; **3,** flower, × 2; **4,** outer sepal, × 4; **5,** inner sepal, × 4; **6,** section of corolla with trifid lobe, outer aspect, × 4; **7,** same, inner aspect, × 4; **8,** section of corolla with trifid lobe, stamen and staminodes, × 4; **9,** stamen and staminodes, outer aspect, × 4; **10,** staminode, × 6; **11,** ovary, × 4; **12,** diagrammatical transverse section of ovary; **13,** fruit, × ⅔; **14,** seeds, × 1. 1, from *Milne-Redhead & Taylor* 7300; 2–14, from *Faulkner* 688.

endosperm absent; cotyledons plano-convex and occupying the whole of the seed.

A genus of at least four species, confined for the present to East and Southern Africa. It is closely related to *Mimusops* but differs in the large coriaceous fruit and seeds with scar covering major proportion of the surface and containing large fleshy cotyledons with endosperm absent. The leaves are characteristically crowded at the shoot tips.

Stipules bristle-like, persistent; leaf-lamina oblanceolate, usually 2–3·2 cm. wide on flowering branches; buds narrowly ovoid or ovoid-ellipsoid, ± acuminate . . . . . 1. *V. kirkii*

Stipules subulate, early caducous; leaf-lamina obovate-elliptic to ± obovate, usually 3–5 cm. wide on flowering branches; buds broadly ovoid and blunt at apex . . . . . 2. *V. cuneata*

1. **V. kirkii** (*Baker*) *Dubard* in Ann. Mus. Col. Marseille, sér. 3, 3: 45 (1915); K.T.S.: 530 (1961). Type: Kenya, Mombasa area, *Kirk* (K, holo.!)

Spreading much-branched shrub or small tree, height up to 5 m. Young shoots with brownish indumentum, later becoming glabrous. Petioles up to 7 mm. long, practically glabrous when mature. Stipules linear, persistent. Leaf-lamina ± oblanceolate, up to 7 cm. long and 3·2 cm. wide, shortly acuminate or rarely ± obtuse, base tapering and narrowly cuneate, practically glabrous; lateral nerves finely raised on lower surface. Pedicels up to 2·8 cm. long, densely ferrugineously pubescent. Outer calyx-lobes narrowly ovate, 6·5–8 mm. long, 2·5–3 mm. wide, with dense ferrugineous pubescence outside; inner lobes oblong-lanceolate, a little smaller and with paler pubescence. Corolla creamish to pale yellow; outer segments ± oblong-lanceolate, up to 7 mm. long; median segment ligulate, erect, 5–7 mm. long; tube up to 2 mm. long. Filaments tapering and ± flattened, up to 2·5 mm. long; staminodes narrowly ovate, up to 4 mm. long, closely adjacent and ± connate at base. Ovary densely pilose; style up to 9 mm. long. Fruit ± ovoid with short terminal rostrum, up to 4 cm. in diameter, indumentum rubbing off and skin becoming practically glabrous. Seeds up to 2·8 cm. in diameter, with lateral hilum and large spreading scar covering over half of surface area. Fig. 12.

KENYA. Mombasa, Port Tudor, *MacNaughton* 85 in *F.D.* 2620!
TANGANYIKA. Pangani District: Bushiri, 6 Oct. 1950 (fl.), *Faulkner* 688! & 2 Dec. 1950 (fl.), *Faulkner* 732!; Uzaramo District: Kiserawe, Hundogo Forest Reserve, Aug. 1953 (fl.), *Paulo* 117!
DISTR. **K7**; **T3**, 6, 8; not known elsewhere
HAB. Lowland dry evergreen forest and coastal evergreen bushlands, usually a marginal component; 0–300 (?–360) m.

SYN. *Butyrospermum ? kirkii* Baker in F.T.A. 3: 505 (1877)
Mimusops bakeri Baill. in Bull. Soc. Linn. Paris 2: 942 (1891) & Hist. Pl. 11: 270 (1892); Engl., E.M. 8: 80 (1904); T.S.K.: 122 (1935). Type: as *Butyrospermum kirkii* Baker *non Mimusops kirkii* Baker

NOTE. A gathering obtained in the Kwale District of Kenya between Samburu and Mackinnon Road, *Drummond & Hemsley* 4203!, may be related to this species. The leaves are shorter in relation to breadth and are generally larger than those of the type, approaching an obovate shape. Flowers are also a little larger in all parts, especially staminodes. Accurate determination must await further collection of material, to show mature fruits and possible leaf shape and size variation. The locality was a small rocky hill at Taru, south of the road in gully shrub-thickets between outcropping rocks.

2. **V. cuneata** (*Engl.*) *Aubrév.* in Adansonia 3: 42 (1963). Types: Tanganyika, W. Usambara Mts., Mashewa, *Holst* 8809 (B, syn. †) & Kwa Mshusa, *Holst* 8976 (B, syn. †, HBG, K, isosyn. !)

Shrub up to 6 m. high with repeated subapical branching; branchlets with pale brownish-grey bark. Petioles 0·5–2 cm. long, practically glabrous. Leaf-lamina obovate-elliptic to ± obovate, (5–)7–11 cm. long, (2·5–)3–5 cm. wide, ± obtuse, cuneate, slightly puberulous along midrib beneath, especially when young; nervation inconspicuous on upper surface, primary lateral nerves ascending and finely raised on lower surface. Flowers clustered in axils of ± terminal leaf-rosettes; pedicels up to 2·5 cm. long, with ± floccose ferrugineous pubescence. Calyx with dense floccose ferrugineous pubescence; outer sepals ± lanceolate, up to 8 mm. long; inner sepals with greyish-brown indumentum. Corolla yellow; segments up to 7 mm. long; tube ± 1·5 mm. long. Filaments up to 2·5 mm. long; staminodes ± ovate, up to 3·5 mm. long, densely woolly externally, ± connivent. Ovary densely pilose; style up to 7 mm. long. Fruits unknown.

TANGANYIKA. Lushoto District: W. Usambara Mts., Lutindi, July 1893 (fl. buds), *Holst* 3420 !

DISTR. **T3**; not known elsewhere

HAB. Bushland and thickets on margins of dry evergreen or deciduous forest; 400–1100 m.

SYN. *Mimusops cuneata* Engl., P.O.A. C: 307 (1895) & E.M. 8: 70, t. 23/C (1904); T.T.C.L.: 565 (1949)
Austromimusops cuneata (Engl.) Meeuse in Bothalia 7: 355 (1960)

NOTE. There seems little doubt that this species is rightly referred to *Vitellariopsis*, with which the arrangement and venation of the leaves and the flowers entirely agree, but the fruits and seeds are still unknown. The plant should be searched for in the drier margins of the NE. side of the W. Usambara Mts. and also on the escarpment below Lutindi at the SW. corner of the same massif. The species seems to be closely related to *V. marginata* (N. E. Br.) Aubrév. from Mozambique to the Cape Province of South Africa. *Drummond & Hemsley* 4203, mentioned under *V. kirkii*, also approaches *V. cuneata* in flower characters, but the persistent stipules and leaf-shape are more like *V. kirkii*.

13. MANILKARA

Adans., Fam. 2: 166 (1763); Dubard in Ann. Mus. Col. Marseille, sér. 3, 3: 6 (1915); H. J. Lam et al. in Blumea 4: 323 (1941), *nom. conserv.*

Eichleria Hartog in J.B. 16: 72 (1878), *nom. illegit.*
Muriea Hartog in J.B. 16: 145 (1878), pro parte; Meeuse in Bothalia 7: 376 (1960); Baehni in Boissiera 11: 86 (1965)
Mahea Pierre, Not. Bot. Sapot.: 8 (1890)
Mimusops L. subgen. *Ternaria* (A. DC.) Engl., E.M. 8: 55 (1904)

Trees or rarely shrubs. Leaves terminal; stipules absent or sometimes present and then soon falling away. Leaf-lamina frequently obovate or elliptic, coriaceous; lower surface with very closely appressed indumentum, giving a silvery light-reflecting effect or subglabrous; midrib prominently raised beneath, primary and secondary lateral nerves closely parallel, venation reticulate. Flowers usually many, clustered in axils of current or recently fallen leaves, long-pedicellate. Sepals 6, arranged in two dissimilar whorls of 3, free or slightly fused. Corolla of 6 members,* fused at base into a short tube, lobes usually divided again into 3 segments. Stamens 6,** epipetalous; anther-dehiscence extrorse. Staminodes 6, alternating with stamens, ± ovate and petaloid to ± ligulate, glabrous, dentate, laciniate or

* See note, p. 2.
**Aberrations of the androecium may occur, see particularly p. 72.

sometimes bifurcate. Ovary with 6–16 locules; ovules solitary, axile, anatropous to campylotropous. Fruit baccate, fleshy or dryish and coriaceous, 1–several-seeded. Seeds ellipsoid to obovoid, laterally compressed; scar basiventral (at least in East Africa), narrowly elliptic to subcircular (not in East Africa); albumen present; cotyledons flattened and foliaceous.

A pantropical genus of about 50 species, with some 15 in Africa, the rest mostly in America and the Pacific Is.

Manilkara zapota (L.) van Royen (*Achras zapota* L. (1753), pro parte, *A zapota* L. var. *zapotilla* Jacq. (1763), *Sapota achras* Mill. (1768), *Manilkara zapotilla* (Jacq.) Gilly (1943), *M. achras* (Mill.) Fosberg (1964), *Nispero achras* (Mill.) Aubrév. (1965)),* native of Central America, has been cultivated at Amani (*Greenway* 929 ! & 2916 !), also in Zanzibar and Pemba (see U.O.P.Z.: 103 (1949)). It bears edible fruit, known as sapodilla or nispero, and the latex is a source of chicle gum used in chewing gum. *M. bidentata* (A. DC.) A. Chev. (*Mimusops balata* auct. *non* Aubl.), a source of balata rubber and a substitute for chicle, is native to Central and South America and has also been tried at Amani (*Greenway* 2801 !).

Leaves in terminal rosettes, mainly on short thick much-scarred lateral branchlets . . . . 1. *M. mochisia*
Leaves ± terminal but not in rosettes:
- Leaves on mature shoots large, mostly not less than 15 cm. in length; primary nerves usually more than 1 cm. apart; vein reticulum conspicuous, especially on lower surface . 2. *M. dawei*
- Leaves smaller, laminae mostly less than 15 cm. in length; primary nerves less than 1 cm. and usually less than 8 mm. apart; veins inconspicuous:
  - Leaves on mature shoots small, mostly less than 4 cm. in length, glabrous beneath . 3. *M. sulcata*
  - Leaves larger, laminae mostly 4·5–15 cm. long, if less then with dense greyish appressed indumentum beneath:
    - Upper surface of leaf with nerves and veins forming a coarse and deeply impressed reticulum . . . . 4. *M. sansibarensis*
    - Upper surface not as above, nerves and veins usually finely raised and inconspicuous:
      - Leaf-lamina elliptic to ± oblanceolate, usually tapering to ± acute or acuminate apex . 5. *M. butugi*
      - Leaf-lamina obovate to elliptic-obovate, apex usually rounded or blunt and emarginate:
        - Leaves beneath appearing ± glabrous even when young, under a hand-lens (hairs present, but imperceptible at this magnification) . . . . . . 6. *M. obovata*
        - Leaves beneath with dense silvery or greyish indumentum of minute appressed hairs:**
          - Leaf-laminae mostly more than 10 cm. long, rarely less and then flower-buds and calyx with dense ferrugineous indumentum . . . . 7. *M. multinervis*

* See H. E. Moore & W. T. Stearn in Taxon 16: 382–395 (1967).
** If reddish and rubbing off on older leaves see *M. sp.* 2, p. 73.

Leaf-laminae mostly less than 10 cm. long, rarely longer and then flower-buds and calyx with brown or greyish-brown indumentum . . . . . . 8. *M. discolor*

1. **M. mochisia** (*Baker*) *Dubard* in Ann. Mus. Col. Marseille, sér. 3, 3: 26 (1915); Meeuse in Bothalia 7: 369, fig. 16 (1960); K.T.S.: 526 (1961); F.F.N.R.: 321 (1962); Meeuse in F.S.A. 26: 50, fig. 8 (1963); J. H. Hemsl. in K.B. 20: 483 (1966). Type: Mozambique, Tete, *Kirk* (K, lecto. !)

Small to medium-sized tree with low branching and spreading crown, height up to 20 m., or sometimes a shrub. Bark brownish-grey or blackish with longitudinal fissures. Branching very irregular with leaves borne mainly in terminal rosettes on dwarf shoots, the latter usually swollen and with rough scarring; young shoots glabrescent or pubescent with evanescent ferrugineous indumentum. Petioles 1·5–12(–20) mm. long, glabrous or puberulous to pubescent especially when young. Leaf-lamina coriaceous, elliptic-obovate to obovate, 1·5–6·5(–7·5) cm. long, 0·8–3 cm. wide, apex rounded, often emarginate, base broadly to narrowly cuneate, in which case usually decurrent with petiole, or rarely ± rounded (especially in very shortly petiolate leaves), glabrous or puberulous to ± pubescent mainly on lower surface especially in young leaves; lateral nerves ± 10–14 on each side, nervation lightly impressed on both surfaces. Flowers white to pale yellow, densely clustered in leaf axils. Pedicels 6–13 mm. long, glabrous or with varying degrees of pubescence. Calyx-lobes ± free to base; outer lobes ± ovate, 2·5–4 mm. long, 1·5–3 mm. wide, glabrescent to densely pubescent; inner lobes similar in shape but slightly smaller and pubescent externally. Corolla-tube very short; lobes trifid; segments narrowly lanceolate or ligulate, 2–4 mm. long. Filaments 1·5–2 mm. long; anthers 1–2 mm. long. Staminodes small, truncate, apex irregularly toothed, rarely with one tooth enlarged and filament-like. Ovary subglobose, shortly pilose, tapering at apex to a simple style, 2–3 mm. long. Fruit yellow when mature with crimson soft edible pulp, subglobose to ellipsoid, up to 1·8 cm. long, 1·3 cm. in diameter, glabrous, containing 1–3 seeds. Seeds dark brown, ellipsoid and compressed, up to 1·3 cm. long, 8 mm. wide; scar lateral, extending to base.

KENYA. Kwale District: between Samburu and Mackinnon Road, near Taru, 9 Sept. 1953, *Drummond & Hemsley* 4217 !; Kilifi District: N. Giriama, Jan. 1937 (fl.), *Dale* in *F.D.* 3663 !; Tana River District: 48 km. on Malindi–Garsen road, 30 Oct. 1961 (yng. fr.), *Polhill & Paulo* 677 !

TANGANYIKA. Tabora, 30 Apr. 1924 (fr.), *Swynnerton* 77 !; Mpwapwa, 18 Oct. 1930 (fl.), *Hornby* 322 !; Morogoro Fuel Reserve, Nov. 1954 (fl.), *Semsei* 1867 !

DISTR. **K**1, 3, 4, 7; **T**1, 3–6, 8; extends northwards to Somali Republic (S.) and southwards through Mozambique, Zambia and Rhodesia to Natal, Botswana and Angola

HAB. Deciduous bushland and thickets, dry scrub with trees; 0–2100 m.

SYN. *Mimusops mochisia* Baker in F.T.A. 3: 506 (1877); Engl., E.M. 8: 63, t. 22/B (1904)

M. densiflora Engl. in P.O.A. C: 307 (1895) & E.M. 8: 63, t. 22/C (1904); T.S.K.: 122 (1936), *non* Baker, *nom. illegit.* Type: Tanganyika, Pangani, *Stuhlmann* coll. 1, 584 (HBG, lecto. !, K, frag.!)

M. densiflora Engl. var. *paolii* Chiov., Coll. Bot. Steph.-Paoli: 111 (1916) & Fl. Somala 2: 274, fig. 157 (1932). Type: Somali Republic (S.), Giuba, Goriei–El Magu, *Paoli* 632 (FI, holo.)

Manilkara densiflora Dale, Addit. and Correct. T.S.K.: 25 (1939); T.T.C.L.: 563 (1949)

For further synonymy and discussion see J. H. Hemsley in K.B. 20: 483—490, fig. 2 (1966)

NOTE. The Chiovenda variety, based on material collected in the Giuba District of Somali Republic (S.), is differentiated on the very small and obscurely toothed staminode. This staminode form has been found in other specimens from NE. Kenya and represents an extreme from the northern end of the distribution range. It merges gradually into the slightly longer and more obviously denticulate forms found further south. Specimens from the central and western regions of Tanganyika possess well-marked pubescence on young leaves, young stems and petioles and on pedicels and flower-buds in contrast to the almost glabrous appearance of the coastal and near-coastal plants. This pubescence seems to be a more prominent feature of Zambian material and appears to follow the north–south inland distribution pattern, no clear-cut limits having been discovered at the present time. It has been regarded by the present author as a variable character within a single species.

See also *M. fischeri*, page 72.

2. **M. dawei** (*Stapf*) *Chiov.* in Atti R. Accad. Ital. 11: 46 (1940), in obs.; I.T.U., ed. 2: 399, fig. 81/a (1952); Verdcourt in K.B. 11: 453 (1957). Type: Uganda, W. Ankole Forest, *Dawe* 353 (K, holo.!)

Small to medium-sized tree, height up to 25 m.; branchlets glabrous; terminal bud and very young shoots with resinous exudate. Petioles robust, 1·5–4 cm. long, glabrous. Leaf-lamina elliptic or elliptic-oblong to obovate-oblong, 10–29 cm. long, 3·5–9·5 cm. wide, coriaceous, apex rounded to emarginate, sometimes shortly acuminate, cuneate; upper surface dark green and glabrous with raised lateral nerves and vein reticulum, lower surface silvery-grey or whitish with indumentum of very small closely appressed regularly arranged hairs, with (16–)18–22 raised and very prominent primary lateral nerves on each side. Young flower-buds with resinous exudate. Flowers fascicled in axils of current or recently fallen leaves, usually 6–8 per axil. Pedicels 2–6 mm. long, stout, with dense pale brown indumentum. Calyx-lobes fused for ± 2 mm. at base; outer lobes elliptic-oblong, up to 8 mm. long, 4 mm. wide, with dense pale silvery-green indumentum externally; inner lobes smaller and less hairy. Corolla white or greenish-white; tube cylindrical, up to 6 mm. long; lobes trifid; outer segments linear-oblong, up to 5 mm. long; median segment spathulate, 6 mm. long. Filaments up to 5 mm. long; anthers up to 2·5 mm. long; staminodes narrowly ligulate, up to 4 mm. long, apex laciniate with 2 (or sometimes more) long delicate processes. Ovary subglobose; style long and tapering, up to 1·4 cm. long, glabrescent; stigma simple. Young fruits with short thick pedicels, subglobose, with very short pale mealy indumentum rubbing off with age; style persistent, conspicuous. Mature fruits not seen.

Uganda. W. Nile District: Payida, 19 Mar. 1945, *Greenway & Eggeling* 7231!; Busoga District: Butembe, Kagoma nursery, 17 Nov. 1950 (fl.), *Osmaston* 443!; Mengo District: near Kampala, Kawanda, Jan. 1936 (yng. fr.), *Chandler* 1558!
Tanganyika. Bukoba District: Kiao I., Sept.-Oct. 1935 (fl.), *Gillman* 390!
Distr. **U**1–4; **T**1; Congo Republic (Lake Kivu area) and Central African Republic
Hab. Lowland rain-forest and riverine forest; 1100–1600 m.

Syn. *Mimusops dawei* Stapf in J.L.S. 37: 523 (1906)
Manilkara aubrevillei R. Sillans in Bull. Soc. Bot. Fr. 99: 42 (1952). Types: Central African Republic [Ubangi-Shari], *Tisserant* 620 & 867 (P, syn.)

3. **M. sulcata** (*Engl.*) *Dubard* in Ann. Mus. Col. Marseille, sér. 3, 3: 26 (1915); T.T.C.L.: 564 (1949); K.T.S.: 527 (1961). Types: Tanganyika, Lushoto District, Mashewa, *Holst* 3551* (B, syn. †, HBG, K, isosyn.!) & Pangani, *Stuhlmann* 176 (B, syn. †, K, isosyn.!)

Small much-branched evergreen tree or shrub, height up to 10 m.; bark grey with small longitudinal fissures. Terminal buds, young shoots and very young leaves with dense ferrugineous pubescence, older leaves and branchlets glabrous. Petioles 3–10(–12) mm. long, glabrous or puberulous. Leaf-lamina

* HBG, K sheets labelled *Holst* 3551a, but no reason to doubt they are parts of the type gathering.

coriaceous, narrowly to broadly obovate or rarely narrowly elliptic, up to 6·5 (rarely –9·5) cm. long and 2·6 (rarely –3·2) cm. wide; upper surface dark shiny green, lower surface paler green, sometimes with slight ferrugineous pubescence near base; nervation inconspicuous. Flowers 2–6. Pedicels 4–7 mm. long, ferrugineously pubescent. Calyx-lobes connate near base; outer lobes ± ovate, 2·5–3·5 mm. long, 1·5–2 mm. wide, ferrugineously pubescent externally; inner lobes narrowly elliptic, 3–4 mm. long, 1–1·5 mm. wide, puberulous externally. Corolla-tube up to 1 mm. long; lobes pale green, ± trifid; two outer segments ± lanceolate, 2·5–4 mm. long; median segment variable in shape, ± ligulate to narrowly elliptic, or rarely very small and subulate or absent, up to 4 mm. long. Filaments 1·5–3 mm. long; anthers ± 1 mm. long. Staminodes irregular in shape, ± triangular to oblong, very small, serrulate, denticulate or ± laciniate, usually with + terminal acumen. Ovary globose, ± 1 mm. long, densely ferrugineous pilose; style up to 3 mm. long. Fruit greenish, ellipsoid to obovoid, up to 1 cm. long, 7 mm. in diameter, subglabrous, 1-seeded. Seed obliquely ovoid, compressed, up to 8 mm. long, 5 mm. in diameter; scar lateral, extending to base.

KENYA. Northern Frontier District: Mararani, Boni Forest, 25 Dec. 1946 (fl.), *Bally* 5988!; Kwale District: 3 km. E. of Mackinnon Road, 9 Sept. 1953 (fr.), *Drummond & Hemsley* 4231!; Kilifi District: Arabuko, May 1929 (fl.), *R. M. Graham* in *F.D.* 2145!

TANGANYIKA. Lushoto District: Mashewa–Magoma road 9 km. SW. of Mashewa, 8 July 1953 (fr.), *Drummond & Hemsley* 3214!; Handeni District: Kideleko, 12 Nov. 1941 (fl.), *Jusuf bin Mohamed* in *F.D.* 8625!; Uzaramo District: Pugu Forest Reserve, 27 Jan. 1938 (fl.), *Wigg* in *F.H.* 1205!

ZANZIBAR. Zanzibar I., Pwani Mchangani, 26 Jan. 1929 (fl.), *Greenway* 1192! & Mazizini, 9 Dec. 1963 (fl.), *Faulkner* 3326!

DISTR. **K1**, 4, 7; **T3**, 6; **Z**; **P**; not known elsewhere

HAB. Lowland dry evergreen forest, coastal woodlands, evergreen bushlands and thickets, also in dry scrub; 0–600(?–1300) m.

SYN. *Mimusops sulcata* Engl., P.O.A. C: 307 (1895) & E.M. 8: 62, t. 22/A (1904); T.S.K.: 122 (1936)
Manilkara sulcata (Engl.) Dubard var. *sacleuxii* Dubard in Ann. Mus. Col. Marseille, sér. 3, 3: 26 (1915). Type: Tanganyika, the area Zanzibar to Bagamoyo District, Mandera, *Sacleux* 993 (P, holo.)

4. **M. sansibarensis** (*Engl.*) *Dubard* in Ann. Mus. Col. Marseille, sér. 3, 3: 26 (1915); T.T.C.L.: 564 (1949); Meeuse in Bothalia 7: 367 (1960); K.T.S.: 526 (1961). Type: Zanzibar I., *Stuhlmann* 1009* (HBG, iso.!)

Small to medium-sized tree with bushy crown, height up to 25 m., with rough greyish or brownish-black bark. Young shoots puberulous or glabrescent. Petioles 9–35 mm. long, glabrous. Leaf-lamina elliptic-obovate to obovate, sometimes ± oblong, up to 14(–15·5) cm. long and 7(–8·8) cm. wide, obtuse and rounded or emarginate at apex, rarely tapering and subacute, cuneate, stiffly coriaceous; upper surface glabrous, nerves and venation impressed, lower surface greyish-green with primary lateral nerves inconspicuous, venation impressed and reticulate, glabrescent or glabrous, sometimes with indumentum of minute closely-appressed hairs.** Flowers fragrant, 4–12 (sometimes fewer), congested in axils of current or recently fallen leaves. Pedicels 5–12 mm. long, densely pubescent. Calyx densely pubescent externally; lobes ± free to base; outer lobes ovate to oblong-ovate, 4–5·5 mm. long, 3–4 mm. wide; inner lobes narrower. Corolla white or greenish-white; tube 1–1·5 mm. long; outer lobes ± lanceolate to

* The type is cited erroneously under *Manilkara densiflora* in the T.T.C.L.: 564 (1949).
** See note on variation below.

narrowly ovate, 3–5 mm. long; median lobe ± elliptic, up to 5 mm. long. Filaments 2·5–3 mm. long; anthers 1·5–2·5 mm. long. Staminodes ± elliptic, narrowly ovate to ± lanceolate, 2·5–3·5 mm. long, margin of distal half irregularly laciniate or serrate. Ovary depressed globose, densely pilose, 10–12(–14)-locular; style 5–9 mm. long. Fruit ellipsoid to subglobose, up to 1·3 cm. long and about the same diameter, indumentum persisting in patches. Seeds 1–4, pale brown, shiny, ± ellipsoid but often straight along one margin, flattened, 7–11 mm. long, up to 6 mm. wide; hilum narrowly obovate, obliquely basal.

Kenya. Kwale District: Mwachi, *Suleman* in *F.D.* 877!; Kilifi District: Mida, Mar. 1930 (fr.), *R. M. Graham* 809 in *F.D.* 2313! & Arabuko, Oct. 1929 (fl.), *R. M. Graham* 695 in *F.D.* 2161!

Tanganyika. Tanga District: Kwale–Tanga, 31 Jan. 1939 (fr.), *Greenway* 5839!; Pangani District: Bushiri, 18 Oct. 1950 (fl.), *Faulkner* 728!; Uzaramo District: Kiserawe, Kurekese Forest Reserve, Sept. 1953 (fl.), *Semsei* 1355!

Zanzibar. Zanzibar I., Sept. 1848 (fl.), *Boivin*! & Mazazini, 4 Nov. 1962 (fl.), *Faulkner* 3125!; Pemba I., Ras Mkumbuu, 22 Dec. 1930, *Greenway* 2768!

Distr. **K**7; **T**3, 6; **Z**; **P**; Mozambique

Hab. Lowland rain-forest and lowland dry evergreen forest, often an important constituent of the latter; also in evergreen woodland and coastal bushlands; 0–300 m.

Syn. *Mimusops annectens* Hartog in J.B. 17: 357 (1879), *nomen subnudum*, based on a *Boivin* specimen from Zanzibar I.
M. sansibarensis Engl., P.O.A. C: 307 (1895) & E.M. 8: 58, t. 21/B (1904)
M. sp. sensu T.S.K.: 90, photo. (1926)
[*M. cuneifolia* sensu T.S.K.: 123 (1936), pro parte maxima,* *nec* Baker, *nec* Dale, Woody Veg. Coast Prov. Kenya (1939)]
[*Manilkara cuneifolia* sensu Dale, Woody Veg. Coast Prov. Kenya: 25 (1939) & sensu Wimbush Cat. Kenya Timbers: 51 (1950), pro parte maxima,* *non* (Baker) Dubard]

Variation. Gatherings from Mafia I. possess a closely appressed indumentum on the lower surface of the leaves which seems to be a constant character from this locality. Similarly the Pemba specimens have appressed hairs on this leaf surface but these vary in density from a uniform and fairly even covering to the occurrence of a few scattered hairs only. The leaves of the Zanzibar specimens are, however, glabrescent and are matched closely by numerous specimens from the mainland.

A specimen from the Uzaramo District of Tanganyika, Mogo Forest Reserve, *Semsei* 1286!, bears small leaves with flower-buds in the axils. Apart from their consistently small size, the leaves of this gathering do not seem to differ in any way from those of *M. sansibarensis*. Dissection of the flower-buds shows the structure to be consistent with that of this species and suggests the plant to be only a minor variant.

5. **M. butugi** *Chiov.* in Atti R. Accad. Ital. 11: 46 (1940); K.T.S.: 525 (1961). Types: Ethiopia, Galla-Sidamo, Dulli Forest, *Giordano* 2465 & Humbi [? Umbi] Forest, *Giugliarelli* 596 (both FI, syn. !)

Tall tree with spreading crown, height up to 35 m., with long straight cylindrical or slightly fluted bole; bark grey-brown, rough and finely fissured. Young shoots subglabrous, usually with pale raised lenticels. Petioles 1–3(–4·3) cm. long, glabrous. Leaf-lamina elliptic to oblanceolate, rarely ± obovate, (5–)8–17(–20) cm. long, (1·6–)3–6·3 cm. wide, upper third of leaf tapering to acuminate apex (especially in crown leaves), rarely acute or ± obtuse, narrowly to broadly cuneate, thinly coriaceous, dull green above, greyish-green beneath with a sparse indumentum of minute closely appressed hairs; lateral nerves ascending, finely raised, looped near margin, intracostal veins present. Flowers 1–6, congested in axils of current leaves. Pedicels stiffly erect or horizontal, 8–12 mm. long, puberulous. Calyx up to 5 mm. long, ± densely pubescent; lobes fused at base forming a cupular structure up to 2 mm. long; outer lobes ovate, up to 3 mm. wide; inner

* The Kakamega tree ludulio (Kak.) is now referred to *Manilkara butugi* Chiov.

lobes slightly smaller. Corolla pale yellow; tube 1·5–2 mm. long; outer lobes narrowly lanceolate, up to 5·5 mm. long; median lobe elliptic, up to 5·5 mm. long. Filaments 2·5–3 mm. long; anthers up to 2 mm. long. Staminodes narrowly oblong to ligulate, up to 2 mm. long, apex with 1–3 (usually 2) slender laciniate processes up to 2 mm. long. Ovary depressed globose, pubescent, 10–12-locular; style 6–8 mm. long, with a base ± swollen and pubescent. Fruits subglobose, up to 3 cm. in diameter, glabrous, with milky pulp. Seed shiny brown, ± obovoid and flattened, 1·3–1·7 cm. long, 7–8·5 mm. wide; hilum lateral, oblique and extending to base.

UGANDA. W. Nile District: near Payida, Feb. 1934 (fl.), *Eggeling* 1512 in *F.D.*1442!; Karamoja District: Timu Forest, June 1946 (fr.), *Eggeling* 5676!; Mbale District: Elgon, Kyosoweri [Kivesoweri], Jan. 1948 (fl.), *Eggeling* 5734!

KENYA. Baringo District: Kabarnet, Apr. 1941 (fr.), *Wimbush* 1217! & 14 June 1957, *Trapnell* 2323!; N. Kavirondo District: Kakamega Forest, June 1933, *Dale* in *F.D.* 3086!

DISTR. U1–3; K2, 3, ? 4, 5; Ethiopia and southern Sudan Republic

HAB. Upland rain-forest, riverine and fringing forest; 1500–2300 m.

SYN. [*Mimusops cuneifolia* sensu Dale, T.S.K.: 123 (1936), pro parte, *non* Baker]
Manilkara sp. sensu I.T.U.: 226, fig. 65/b (1940) & ed. 2: 399, fig. 81/b (1952), based on *Eggeling* 1512 & 3163
[*M. multinervis* sensu I.T.U., ed. 2: 399 (1952), *non* (Baker) Dubard]
Mimusops sp. sensu I.T.U., ed. 2: 402 (1952)
M. sp. sensu Wimbush, Cat. Kenya Timbers: 53 (1956), pro max. parte, excl. nom. mugambwa (Kik.)

NOTE. This tree, in general outline, bears a superficial resemblance to *Mimusops bagshawei*. This is especially marked in the leaf-shape, but the two may be distinguished by the following leaf-characters. In *M. bagshawei* the midrib is raised on the upper surface, the lower surface is glabrescent with no suggestion of a silvery-grey appearance and the vein reticulation is apparent to the naked eye. In *Manilkara butugi* the midrib is immersed in a channel-like fold, the lower surface has a covering of minute appressed hairs imparting a greyish or a silvery-green colour and the vein reticulum is very fine and hair-like and is not easily visible to the naked eye.

6. **M. obovata** (*Sabine & G. Don*) *J. H. Hemsl.* in K.B. 17: 171 (1963); Heine in F.W.T.A., ed. 2, 2: 20 (1963); J. H. Hemsl. in K.B. 20: 468 (1966). Type: Sierra Leone, *G. Don* (BM, holo.!)

Medium-sized to tall tree with long clean and slightly fluted bole and buttressed base, height up to 35 m.; bark pale grey to dark brownish-grey, fissured and rough. Young branchlets (from crown) deep purplish-brown with pale lenticels, glabrous. Scars of fallen leaves prominent, sometimes raised and peg-like. Leaves clustered towards apices of branchlets. Petioles 5–18 mm. long, glabrescent. Leaf-lamina* thinly to thickly coriaceous, obovate (rarely obovate-oblong), 3–10(–17) cm. long, 1·6–5·5(–7·2) cm. wide, apex rounded or emarginate, rarely subacute, cuneate; upper surface fresh green, glabrous, lower surface paler, drying to a brownish-grey, appearing glabrescent;** primary lateral nerves raised, arcuate, very fine and hair-like. Flowers fascicled in axils of older or fallen leaves. Pedicels 4–10 mm. long, with appressed brownish hairs. Calyx-lobes slightly connate at base; outer lobes ± ovate to ovate-oblong, up to 6 mm. long and 3·5 mm. wide, with brownish pubescence externally; inner lobes slightly narrower and with little pubescence externally. Corolla white; tube ± 1·5 mm. long; lobes trifid; outer segments linear-lanceolate, up to 5 mm. long; median segment narrowly elliptic, up to 4·5 mm. long. Stamen-filaments up to 3·5 mm. long, flattened; anthers up to 2·5 mm. long; staminodes oblong to ± linear,

* See note at end of species.

** At least under a hand-lens, magnifications of × 30 or more will show the presence of densely arranged minute appressed hairs covering the entire surface.

irregularly laciniate. Ovary subglobose, densely pilose, 9–11-locular; style up to 6 mm. long; stigma minutely papillate. Fruits incompletely known, yellow when ripe, said to be edible, obovoid to subglobose, up to 2·5 cm. long. Mature seed not known.

UGANDA. Masaka District: Malabigambo Forest near Katera, 14 Aug. 1950, *Dawkins* 615! & S. Buddu, *Dawe* 334! & Kyebe, Aug. 1954 (fl.), *Purseglove* 1786!
TANGANYIKA. Bukoba District: Minziro Forest, Iara, Aug. 1939 (yng. fr.), *Wigg* in *F.H.* 1506! & Minziro Forest, Feb. 1958 (fl.), *Procter* 828!; Musoma District: Buhemba [Ruhemba] mine, *Grant*!
DISTR. **U**4; **T**1; extends from Sierra Leone in the west, through Liberia, Ivory Coast, Ghana and Dahomey, Nigeria, Cameroun Republic, Gabon and Congo Republic to Zambia and Angola
HAB. Lowland rain-forest, riverine forest and swamp-forests of the SW. shores of Lake Victoria, commonly found as a co-dominant canopy tree in the latter with *Podocarpus usambarensis* var. *dawei*, *P. milanjianus* and *Baikiaea insignis* subsp. *minor*; 1100–1300 m.

SYN. *Chrysophyllum obovatum* Sabine & G. Don in Trans. Hort. Soc. Lond. 5: 458 (1824)
Mimusops cuneifolia Baker in F.T.A. 3: 506 (1877); Engl., E.M. 8: 64 (1904). Type: Angola, Cabinda Province, *C. Smith* (K, holo.!)
M. lacera Baker in F.T.A. 3: 507 (1877); Engl., E.M. 8: 59, t. 20/B (1904). Types: Nigeria, Nupe, *Barter* 1270 & Nun R., *Mann* 489 (both K, syn.!)
M. welwitschii Engl. in E.J. 12: 524 (1890) & E.M. 8: 58 (1904). Type: Angola, Cuanza Norte, Queta, *Welwitsch* 4814 (K, iso.!)
M. propinqua S. Moore in J.L.S. 37: 177 (1905). Type: Uganda, Masaka District, Musozi, *Bagshawe* 76 (BM, holo.!, K, iso.!)
Chrysophyllum holtzii Engl. in E.J. 49: 390 (1913); T.T.C.L.: 562 (1949). Type: Tanganyika, Bukoba District, Minziro Forest, *Holtz* 1697 (B, holo. †)
Manilkara cuneifolia (Baker) Dubard in Ann. Mus. Col. Marseille, sér. 3, 3: 23, fig. 11 (1915); I.T.U., ed. 2: 396, fig. 82/b, c (1952)
M. lacera (Baker) Dubard in Ann. Mus. Col. Marseille, sér. 3, 3: 24 (1915); Aubrév., Fl. For. Côte d'Ivoire 3: 98, t. 279/1–7 (1936) & Not. Bot. Syst. 16: 227 (1960) & Fl. Gabon 1: 31 (1961) & Fl. Cameroun 2: 31, t. 2/4–8 (1964)
M. welwitschii (Engl.) Dubard in Ann. Mus. Col. Marseille, sér. 3, 3: 27 (1915)
M. propinqua (S. Moore) H. J. Lam in Blumea 4: 356 (1941)
M. sp. 1 & *M. sp. 2* sensu F.F.N.R.: 321 (1962)

NOTE. Leaves from saplings may have an abruptly tapered acumen up to 1 cm. long and in addition the lower surface may have a greyish silky appearance due to very small and densely arranged appressed hairs. Such leaves may resemble those of *Bequaertiodendron natalense*.

The species as here accepted is widespread and includes a number of smaller geographical segregates. Characters such as degree of laciniation and number of segments of staminodes, and minor differences in leaf-size, which previously have been used to distinguish species, have proved impossible to work within the range of material now available.

The tree yields a useful timber, known as Nkunya in Uganda, which is very hard and durable in the ground (see Uganda For. Dept. Timber Leaflet No. 3). A single sterile gathering from the E. Usambara Mts. may possibly extend the species range into E. Tanganyika (see under imperfectly known species No. 1, page 73).

7. **M. multinervis** (*Baker*) *Dubard* in Ann. Mus. Col. Marseille, sér. 3, 3: 24 (1915); Aubrév., Fl. For. Soud.-Guin.: 425, t. 93/2 (1950) & Fl. For. Côte d'Ivoire, ed. 2, 3: 120, t. 293/8 (1959) & Not. Syst. 16: 227, fig. 1/1–4 (1960); Heine in F.W.T.A., ed. 2, 2: 20 (1963); Fl. Cameroun 2: 28, t. 1/1–4 (1964); Ic. Pl. Afr. 6, t. 141 (1964); J. H. Hemsl. in K.B. 20: 490 (1966). Type: N. Nigeria, Nupe, *Barter* 1123 (K, holo.!)

Small to medium-sized tree, height up to 6–27 m.; bark rough and fissured. Shoot-tips brown pubescent, soon glabrescent. Stipules linear-subulate, up to 6–8 mm. long, caducous. Petioles 1·5–4 cm. long. Leaf-lamina elliptic-oblong to elliptic-obovate, up to 9–15(–19) cm. long, 3·5–7(–9·5) cm. wide, apex rounded and usually shortly acuminate, narrowly to broadly cuneate, coriaceous; upper surface glabrous, lower surface with a dense very short appressed silvery indumentum or (outside East Africa) appearing glabrous, though still with minute hairs present; primary lateral nerves numerous, slightly raised and very fine. Flowers clustered in axils of older leaves and at the nodes below; pedicels 7–10 mm. long. Calyx-lobes almost free to the base; outer lobes ovate to elliptic, 3·5–5·5 mm. long, with a dense short brown pubescence externally. Corolla creamy-white; tube (0·5–)0·8–1 mm. long; lobes trifid; outer segments lanceolate, 3–4 mm. long; median segment elliptic-oblong and of similar length. Filaments up to 1·5–2 mm. long; anthers 1·5–1·7 mm. long; staminodes oblong, ± 1–1·3 mm. long, with 2–3 slender sometimes laciniate 1–2 mm. long processes at the apex. Ovary depressed-globose, pubescent, 9–16-locular; style 3–4·5 mm. long, tapered. Fruit ripens yellow, subglobose to slightly obovoid, 1·5–2 cm. long, ultimately glabrescent, 1–several-seeded. Seed light-brown, obliquely obovoid and flattened, basally pointed, ± 11–12 mm. long, 7 mm. wide; scar narrow, lateral and oblique, extending little over half seed-length.

subsp. **schweinfurthii** (*Engl.*) *J. H. Hemsl.* in K.B. 20: 497 (1966). Types: Sudan Republic, Equatoria Province, *Schweinfurth* 1378 & 1529 (both B, syn. †, K, isosyn. !) & 1777 (B, syn. †)

Lower surface of leaves with soft silky indumentum of minute densely arranged hairs visible with a hand-lens; primary lateral nerves arcuate, sometimes obscurely looped at tips, vein network inconspicuous; ovary 9–14-locular.

UGANDA. W. Nile District: Moyo–Arua, *Dale* 812 ! & Moyo, 26 Nov. 1941 (fl. bud), *A. S. Thomas* 3752 !; Acholi District: Gulu–Kitgum road by Aswa R., *Eggeling* 777 in *F.D.* 1163 !

DISTR. **U1**; southern Sudan Republic and NE. Congo Republic to Central African Republic

HAB. Riverine forest and open deciduous woodland; 900–1050 m.

SYN. *Chrysophyllum sp.* sensu Baker in F.T.A. 3: 501 (1877), quoad *Grant* 703
Mimusops schweinfurthii Engl. in E.J. 12: 523 (1890) & E.M. 8: 87, t. 20/D (1904)
Manilkara schweinfurthii (Engl.) Dubard in Ann. Mus. Col. Marseille, sér. 3, 3: 25 (1915); I.T.U., ed. 2: 399, fig. 82/a (1952); F.P.S. 2: 374 (1952)

NOTE. Subsp. *multinervis* replaces subsp. *schweinfurthii* west from Central African and Cameroun Republics to Mali and Guinée Republic. It differs in that the lower leaf-surface appears glabrous under a hand-lens though in fact covered with very minute hairs, the primary nerves are more distinct, finely raised and looped at the tips, and it has a 12–16-locular ovary.

8. **M. discolor** (*Sond.*) *J. H. Hemsl.* in K.B. 20: 510 (1966); K.T.S.: 526 (1961). Types: South Africa, Natal, Durban, *Gueinzius* 128 & 547 (S, syn.)

Small to medium tree, height up to 30 m., with rough dark grey or blackish bark. Terminal buds, young shoots and petioles glabrescent. Petioles 0·5–1·7 cm. long. Leaf-lamina oblong-elliptic to oblong-obovate, up to 9·5(–11) cm. long, 4·5(–5·3) cm. wide, apex rounded and slightly emarginate or rarely shortly acuminate, broadly to narrowly cuneate, coriaceous; upper surface deep green with reticulum of very fine veins, lower surface silvery-grey with dense indumentum of minute closely appressed regularly arranged hairs sometimes brownish on midrib and main nerves or midrib glabrescent. Flowers axillary, usually in clusters of 4–6 in axils of current leaves. Pedicels 3–10 mm. long, with brown pubescence. Calyx-lobes connate near base; outer lobes ovate or ovate-oblong, 4·5–6 mm. long, 3–3·7 mm. wide, with brown pubescence externally; inner lobes slightly smaller and with less pubescence. Corolla yellow; tube ± 1·5 mm. long; lobes trifid, or entire in ♀ flowers; two outer segments ± lanceolate, up to 4 mm. long; median segment ± elliptic, up to 5 mm. long, tapering to narrow basal attachment.

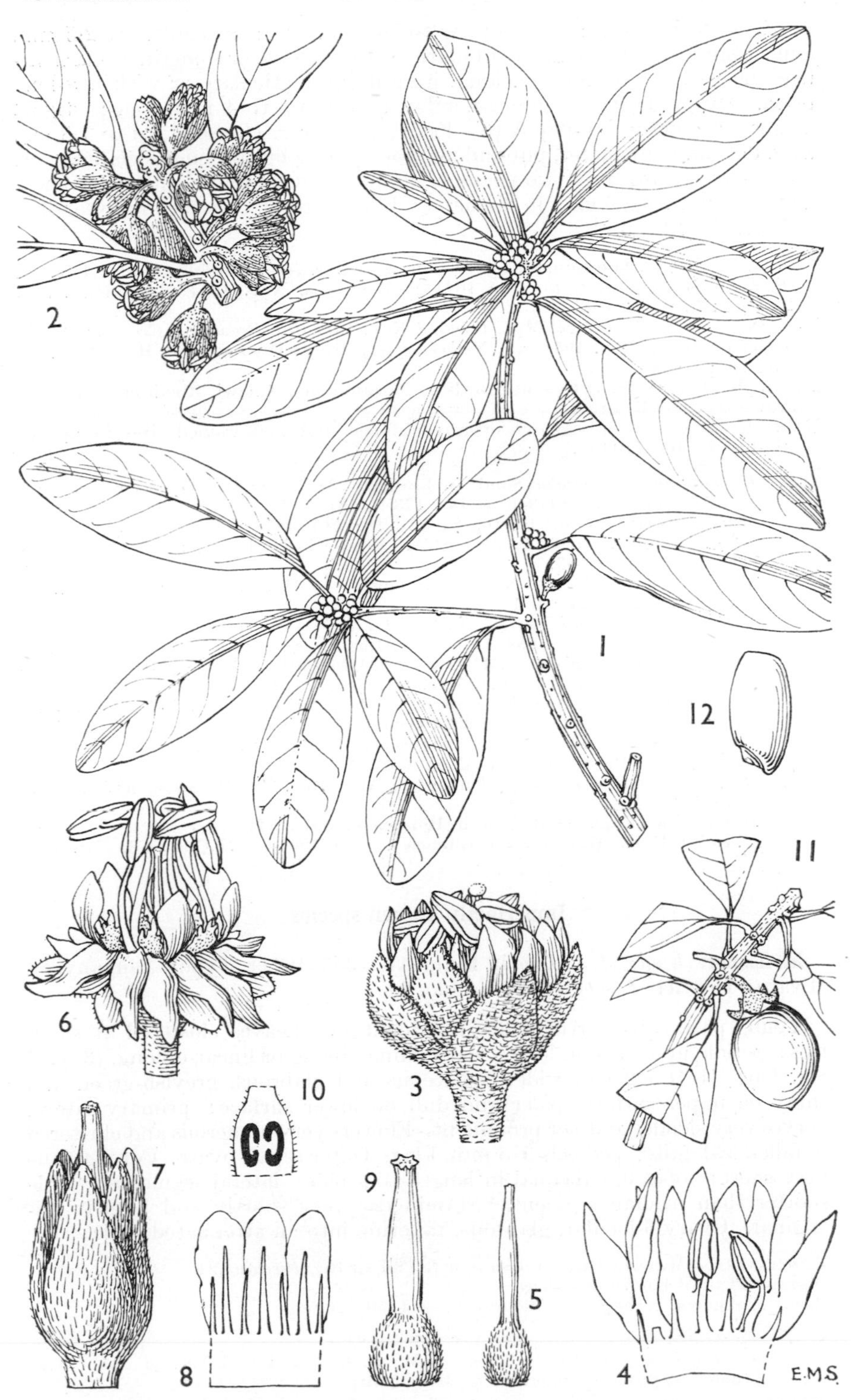

FIG. 13. *MANILKARA DISCOLOR*—**1,** branch, × $\frac{2}{3}$; **2,** inflorescences, × 1; **3,** flower, × 4; **4,** corolla-section of same, × 4; **5,** ovary of same, × 4; **6,** more mature flower, × 4; **7,** reduced female flower, × 8; **8,** corolla-section of same, × 8; **9,** ovary of same, × 8; **10,** section of ovary, × 8; **11,** young fruit, attached, × 1; **12,** seed, × 2. 1, from *G. R. Williams* 420; 2–5, from *Gardner* in *F.D.* 1024; 6, from *G. R. Williams* 663; 7–10, from *Timothy* 548 in *C.M.* 15954; 11, 12, from *Nicholson* 49.

Fertile stamens 6–12*; filaments up to 3·5 mm. long; anthers up to 2·5 mm. long. Staminodes truncate or subulate, ± 1 mm. long, or sometimes filament-like and then up to 3·5 mm. long, irregularly denticulate or with few long teeth. Ovary depressed globose, pilose; style up to 4 mm. long; stigma simple or sometimes 6-papillate. Fruit ovoid or ± ellipsoid, up to 1·3 cm. long, 0·8 cm. in diameter, puberulous when young but becoming glabrescent. Seeds solitary or rarely two present, obliquely ovoid and slightly flattened, up to 10 mm. long, 6 mm. in diameter; hilum lateral, oblique and extending to base. Fig. 13, p. 71.

KENYA. Nairobi Arboretum, 2 June 1952 (fl. & yng. fr.), *G. R. Williams* 420!; Kiambu District: Ngong Forest, *Gardner* 1024!; Machakos, Nov. 1954 (fl. buds), *D. G. Leakey* in *E.A.H.* 270/54!

TANGANYIKA. W. Usambara Mts., Kwai, 8 Nov. 1947 (fl.), *Brenan & Greenway* 8310! & Mlola road, 18 Nov. 1954 (fl.), *Nicholson* 6!; Iringa District: Sao Hill, Oct. 1935 (fl.), *Hornby* 658!

DISTR. **K** ?3, 4–6; **T**3, 7, 8; southwards to Mozambique, Malawi, Rhodesia and Natal Province of South Africa

HAB. Lowland and upland dry evergreen forest and well-drained sites in upland rain-forest; 400–2100 m.

SYN. *Labourdonnaisia discolor* Sond. in Linnaea 23: 73 (1850); Gerstner in Journ. S. Afr. Bot. 12: 49 (1946) & 14: 173 (1948)
Muriea discolor (Sond.) Hartog in J.B. 16: 145 (1878); Dubard in Ann. Mus. Col. Marseille, sér. 3, 3: 29 (1915); Meeuse in Bothalia 7: 377 (1960) & in F.S.A. 26: 52 (1963)
Mahea natalensis Pierre, Not. Bot. Sapot.: 10 (1890). Type: South Africa, Natal, no specimen cited
Mimusops buchananii Engl., P.O.A. C: 307 (1895). Type: Malawi, Shire Highlands, *Buchanan* 684 (B, holo. †, BM, K, iso.!)
M. discolor (Sond.) Engl., E.M. 8: 55, t. 34/A (1904)
M. altissima Engl., E.M. 8: 55 (1904); T.T.C.L.: 564 (1949). Type: Tanganyika, Lindi District, Kwa Sikumbi, *Busse* 2905 (B, holo. †, BM, EA, iso.!)
M. eickii Engl., E.M. 8: 60, t. 23/A (1904). Types: Tanganyika, W. Usambara Mts., *Eick* 18 & 42 & *Stuhlmann & Engler* 1246 (all B, syn. †)
M. natalensis (Pierre) Engl., E.M. 8: 65, t. 25/B (1904)
Manilkara natalensis (Pierre) Dubard in Ann. Mus. Col. Marseille, sér. 3, 3: 28 (1915)
M. altissima (Engl.) H. J. Lam in Blumea 4: 354 (1941)
M. eickii (Engl.) H. J. Lam in Blumea 4: 355 (1941); T.T.C.L.: 564 (1949)

Imperfectly known species

M. fischeri (*Engl.*) *H. J. Lam* in Blumea 4: 355 (1941). Type: Tanganyika, Mwanza District, *Fischer* 424 (B, holo. †)

Young parts with ferrugineous indumentum. Leaves clustered at shoot tips; petiole up to 7 mm. long; leaf-lamina oblong or linear-oblong, (3–)6–7 cm. long, (1–)1·5–2 cm. wide, coriaceous and glabrous, greyish-green and shiny on upper surface, paler and dull on lower surface; primary lateral nerves very slender and not prominent. Flowers very numerous and clustered in fallen leaf axils; pedicels 3–5 mm. long. Outer sepals ovate. Corolla-tube very short; lobes 6, subequal in length to sepals; lateral segments a little shorter than median segment. Staminodes very shortly and bluntly tri-angular. Ovary 6-locular, glabrous, tapering into an attenuated style.

TANGANYIKA. Mwanza District: Simiyu R. [Simiu R.], *Fischer* 424
DISTR. **T**1; not known elsewhere
HAB. Unknown, perhaps riverine forest; ± 1150–1200 m.

SYN. *Sideroxylon fischeri* Engl., P.O.A. C: 306 (1895)
Mimusops fischeri (Engl.) Engl., E.M. 8: 64 (1904); T.T.C.L.: 564 (1949) [noted as probably referable to *Manilkara*]

* See K.B. 20: 502–10 (1966).

NOTE. From the description this species appears to be very similar to *M. mochisia*, as pointed out in K.B. 20: 488 (1966), but more material is needed from the type-locality before any firm decision can be made in the absence of authentic material.

M. sp. 1

Medium-sized tree, 25 m. high with much branched crown and yellowish-grey finely fissured bark. Young shoots dark reddish-brown, glabrescent. Petioles 1·5–2·5 cm. long, flattened, narrowly channelled above, glabrous. Leaf-lamina obovate, 7–14 cm. long, 3·4–7·5 cm. wide, very shortly acuminate to rounded, cuneate; upper surface glabrous, midrib impressed, nerves very inconspicuous, lower surface pale brown (dried specimens) and with indumentum of closely appressed minute hairs; primary lateral nerves raised, ascending, looped near margin.

TANGANYIKA. Lushoto District: Ngua–Kwamkoro, 31 Dec. 1936, *Greenway* 4817!
DISTR. T3; known only from this area
HAB. Lowland rain-forest, very rare, with *Cephalosphaera*, *Polyalthia*, *Allanblackia*, *Parinari*; 1000 m.

NOTE. This specimen resembles *M. obovata* (Sabine & G. Don) J. H. Hemsl. in leaf-characters and differs only in the more acutely ascending and less sharply defined lateral nerves. There is no record of *M. obovata* ever having been introduced at Amani, and the occurrence of the present tree in what appears to be climax lowland forest is against the theory of an introduced species. Additional gatherings with flowers and fruits are necessary in order to ascertain the status of this tree.

M. sp. 2

Medium-sized tree, 25 m. high; bark of smaller branches dark grey with fine longitudinal striation. Terminal buds, young shoots and young petioles with ferrugineous pubescence. Petioles 4–12 mm. long. Leaf-lamina elliptic to elliptic-obovate, up to 6 cm. long, 3 cm. wide, apex rounded or slightly emarginate, cuneate, coriaceous; upper surface glabrous, nervation inconspicuous, slightly impressed, lower surface with reddish-brown pubescence of dense small hairs, rubbing off on older leaves; primary nerves inconspicuous. Pedicel 4 mm. long. Outer sepals 4 mm. long, 3 mm. wide, with dense deep brown pubescence; inner sepals slightly smaller and with pale brown pubescence. Ovary densely brown pilose; style 4 mm. long.

TANGANYIKA. Mpanda District: 35 km. E. of Uruwira [Uluwira] on Nyonga road, 3 July 1949, *Wigg* in *F.H.* 2756! & *Hoyle* 1064!
DISTR. T4; Congo Republic
HAB. Riverine forest and locally common in moist sites in *Julbernardia-Brachystegia-Pterocarpus* woodland; ± 1200 m.

NOTE. This material consists of leafy twigs and two old flowers, from which the corollas have dropped. The leaves are not unlike those of *M. discolor* (Sond.) J. H. Hemsl. in general outline, but possess a thick red-brown indumentum in the younger stages. Coupled with a differing colour and texture of the bark and a pronounced habitat difference it is possible that these specimens may represent an undescribed species. One further specimen from Katanga, *Ringoet* 27!, should also probably be included here.

IMPERFECTLY KNOWN GENUS

Small to medium tree, with spreading crown, height up to 20 m.; bark dark brown. Young shoots, buds and young petioles with ferrugineous pubescence of appressed hairs; older branches, petioles and leaves glabrous. Stipules absent. Petioles 3–12 mm. long; leaf-lamina ± ovate to elliptic-ovate or ovate-lanceolate, 2·5–6·5(–8·5) cm. long, 1·3–3·3(–4) cm. wide, coriaceous, acute to ± acuminate, cuneate; primary lateral nerves few, 4–7(–9) each side, slightly impressed and inconspicuous on upper surface, raised on lower surface, arcuate, ascending, secondary nerves rarely present, veins inconspicuous. Flowers solitary or up to 3 in current leaf axils; pedicels ± straight or curved and deflexing, 6–13 mm. long, glabrous. Sepals 5, slightly fused near base, suborbicular, up to 2·5 mm. long, glabrous. Corolla greenish-cream; tube short and open, up to 1·5 mm. long; lobes stiff and fleshy, valvate, rounded to broadly ovate, up to 2·5 mm. long. Filaments short and tapering up to 1 mm. long; anthers ± obcordate; staminodes usually present, ± subulate, up to 1 mm. long. Ovary subglobose, up to 2 mm. in diameter, glabrous; style very short and thick, truncate and with 5–6 pits on flattened stigmatic surface.

TANGANYIKA. W. Usambara Mts., Shagai Forest near Sunga, 17 May 1953 (fl.), *Drummond & Hemsley* 2599! & May 1953 (fl.), *Procter* 225!
DISTR. T3; not known elsewhere
HAB. Upland evergreen forest with *Podocarpus* spp., *Ocotea usambarensis*, *Aningeria adolfi-friedericii* subsp. *usambarensis*, *Chrysophyllum gorungosanum*, etc.; ± 2000 m.

NOTE. This interesting tree has no close affinity with any known African sapotaceous genus. The smallish leaves with few nerves, coupled with a general twigginess of growth and solitary or very few flowers per node readily distinguish it from the other genera. Flower-characters would seem to suggest some relationship with the genus *Sideroxylon*, but it remains for fruits to be obtained before any firm decision can be taken. Collectors, especially local foresters, are requested to obtain fruiting material. The tree gives the pinkish to reddish slash, exuding white latex, typical of the family; the gathering *Drummond & Hemsley* 2599 was obtained about 2 km. E. of the Shagai sawmills.

SPECIES OF DOUBTFUL POSITION

Synsepalum ulugurense (*Engl.*) *Engl.*, E.M. 8: 32 (1904); T.T.C.L.: 568 (1949). Types: Tanganyika, Uluguru Mts., Ng'hweme, *Stuhlmann* 8789 & 8827 (B, syn. †)

Tree with leaves clustered at apices of slender shoots; branches, together with petiole and midrib, densely rusty hairy when young. Stipules present on young shoots, later deciduous, narrowly lanceolate. Petioles 3–5 mm. long. Leaf-lamina narrowly lanceolate-oblong or lanceolate, 12–16 cm. long, 3–4 cm. wide, subcoriaceous, glabrous except on midrib, apex with long acumen to 1·5 cm. long and 3 mm. wide, tapering to narrow base; primary lateral nerves ± 12 pairs, spaced 5–7 mm. apart, prominent on lower surface; veins oblique, very numerous and prominent beneath.

TANGANYIKA. Morogoro District: Uluguru Mts., Ng'hweme, Oct. 1894, *Stuhlmann* 8789 & 8827
DISTR. T6; not known elsewhere
HAB. Upland rain-forest; 1500–1700 m.

SYN. *Chrysophyllum ulugurense* Engl. in E.J. 28: 448 (1900)
Pouteria ? ulugurensis (Engl.) Baehni in Candollea 9: 417 (1942)

NOTE. The above account is taken from the Engler description, no material having been seen. The species is based on gatherings of leafy shoots, without flowers or fruits, and it has not been possible to determine the genus or validity of the species, with any certainty, at the present time; but see also note under *Pachystela subverticillata* E. A. Bruce (page 39).

INDEX TO SAPOTACEAE

GEOGRAPHICAL DIVISIONS OF THE FLORA

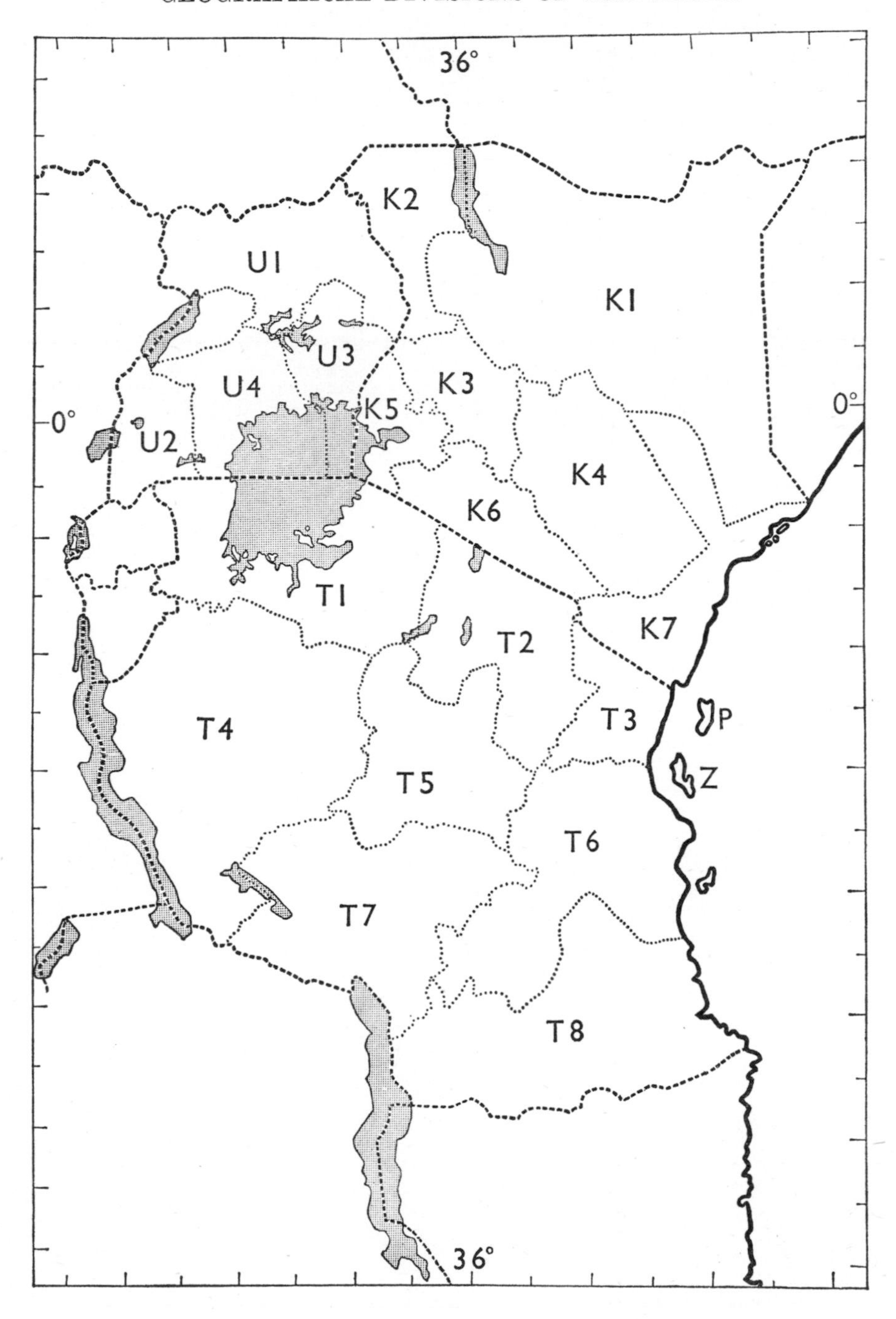